VIRUS INFECTIONS

MICROBIOLOGY SERIES

Series Editors

ALLEN I. LASKIN
EXXON Research and Engineering Company
Linden, New Jersey

RICHARD I. MATELES
Stauffer Chemical Company
Westport, Connecticut

Volume 1 Bacterial Membranes and Walls
edited by Loretta Leive

Volume 2 Eucaryotic Microbes as Model Developmental Systems
edited by Danton H. O'Day and Paul A. Horgen

Volume 3 Microorganisms and Minerals
edited by Eugene D. Weinberg

Volume 4 Bacterial Transport
edited by Barry P. Rosen

Volume 5 Microbial Testers: Probing Carcinogenesis
edited by I. Cecil Felkner

Volume 6 Virus Infections: Modern Concepts and Status
edited by Lloyd C. Olson

Other Volumes in Preparation

VIRUS INFECTIONS

Modern Concepts and Status

Edited by Lloyd C. Olson
The Children's Mercy Hospital
University of Missouri–Kansas City
Kansas City, Missouri

MARCEL DEKKER, INC. New York • Basel

Library of Congress Cataloging in Publication Data
Main entry under title:

Virus infections.

(Microbiology series ; v. 6)
Includes bibliographies and index.
1. Virus diseases. I. Olson, Lloyd C., 1935-
II. Series. [DNLM: 1. Virus diseases. 2. Interferon.
3. Hepatitis viruses. 4. Oncogenic viruses. 5. Viruses.
W1 MI292S v.6 / WC 500 V8153]
RC114.5.V49 616.9'2507 82-5097
ISBN 0-8247-1859-3 AACR2

MARCEL DEKKER, INC.

270 Madison Avenue, New York, New York 10016

Current printing (last digit):
10 9 8 7 6 5 4 3 2 1

PRINTED IN THE UNITED STATES OF AMERICA

Preface

At the midpoint of this century, viruses were invisible organisms which were responsible for certain acute human disease, which killed laboratory mice, and which parasitized bacteria. Conceptually they were elemental parasites which invaded, wreaked their consequences, and were in turn eliminated. Now that is all changed, not only as a consequence of a more thorough understanding of what viruses are, but also because of our more complete appreciation of the nature of the host cells themselves. Numerous viruses have been isolated and characterized, and each has been classified according to its genome, virion structure, relation to its hosts and host cells, and antigenic composition. As knowledge of each becomes more complete, that knowledge becomes more complex and extensively documented.

The techniques employed by investigators to pursue this knowledge have also become more sophisticated, innovative, and ingenious and borrow not only from cellular and molecular biology but also biophysics and electronic engineering. As a result, the interpretation of much of experimental virology has passed outside the immediate concerns of many of those with at least a passing interest in viruses. This is particularly unfortunate in that new viruses continue to be isolated, and the biology of virus infections continues to expand.

To be sure, virology is periodically reviewed. Such reviews, however, tend to be exhaustive and heavily referenced compendia appropriate to a current analysis of the subject. As such, on the other hand, they tend to lack the elements of broad interpretations or speculations necessary to a reasonably durable perspective of the topic. The present volume attempts to provide these elements. An overview of viruses, their relationships with the host cell and the molecular pathogenesis of their effects, is presented first. Next is analyzed the host immune response and how this may lead to an immunopathologic process. One part of this immune response is interferon production, which forms the subject of a detailed review. Finally, our current understanding of the new hepatitis viruses, of the fascinating agents best designated as unconventional viruses, and of the two major groups of tumor viruses, each receive detailed consideration. All discussions are intended to be informative and interpretive rather than a technical reference source. The authors have intended their efforts to provide insight into the intriguing relationships between cell and virus in an entertaining manner. We hope the reader will find it so.

Lloyd C. Olson

Contributors

Daniel W. Bradley U.S. Public Health Service, Department of Health and Human Services, Hepatitis and Viral Enteritis Division, Centers for Disease Control, Phoenix, Arizona

Erik D. A. De Clercq Department of Human Biology, Division of Microbiology, Rega Institute for Medical Research, Katholieke Universiteit Leuven, Leuven, Belgium

Raymond V. Gilden Biological Carcinogenesis Program, NCI-Frederick Cancer Research Facility, Frederick, Maryland

Scott B. Halstead Department of Tropical Medicine and Medical Microbiology, John A. Burns School of Medicine, University of Hawaii at Manoa, Honolulu, Hawaii

David T. Kingsbury* Department of Microbiology, College of Medicine, University of California, Irvine, Irvine, California

Robert M. McAllister Department of Pediatrics, Children's Hospital of Los Angeles, University of Southern California School of Medicine, Los Angeles, California

James E. Maynard U.S. Public Health Service, Department of Health and Human Services, Hepatitis and Viral Enteritis Division, Centers for Disease Control, Phoenix, Arizona

Lloyd C. Olson Department of Infectious Diseases, The Children's Mercy Hospital, University of Missouri - Kansas City, Kansas City, Missouri

**Present affiliation* School of Public Health, University of California, Berkeley, Naval Biosciences Laboratory, Naval Supply Center, Oakland, California.

Contents

VIRUS INFECTIONS

chapter 1

Virus Replication and the Pathogenesis of Disease

LLOYD C. OLSON

The Children's Mercy Hospital
University of Missouri-Kansas City
Kansas City, Missouri

Eukaryotic cells represent the culmination of an immense evolutionary effort directed toward the successful perpetuation of complex life forms. Central to this success is the development of nucleic acid-based systems for storage of information necessary to the proper construction and function of organisms. Perhaps even more remarkable, however, are the systems ensuring coordinated expression of this information in a manner that best serves the individual cell and the total organism. This is especially true when one considers the enormity of the task of starting with an individual array of data to construct such complex multicellular organisms as dinosaurs or elephants. The regulatory mechanisms responsible for expression of information at the proper time, in the proper sequence, and in the proper proportion are complex and exquisitely controlled.

This elegant biosynthetic machinery is also susceptible to expropriation by viruses. That this is a common event in the natural biology of the cell is attested to by the myriad successful schemes pursued by viruses for their own purposes, and by the fact that few if any cells are exempt from virus predations. Intracellular competitions between invading viruses and the host cell represent the fundamental element of the virus-cell relationship. In fact, viruses represent the most basic form of parasitism; they are in essence genetic interlopers. Some viruses interject in place of the cellular chromosome their own DNA message, which, once innocently processed by the cell, provides all the elements necessary to the construction of virus progeny. Other kinds of viruses accomplish the same goal by representing themselves to the cell as the intermediate messenger normally employed for information transfer.

As with all life forms, the sole purpose of a virus is to ensure reproduction in kind. To do so, however, has required that viruses be constructed in a manner that allows progeny virions to escape from the host cell, pass through a hostile extracellular and extraorganismic environment, successfully encounter new host cells and fresh hosts, and recognize and then enter such cells as are capable of supporting renewed replication. The adaptive strategies various viruses have developed for each step of the infectious cycle are diverse; many are fascinating; and were viruses granted the attribute of conscious intent many would best be described as exceptionally ingenious.

In actual fact viruses do not merely provide the genome for the cell's synthesizing machinery without themselves exerting controls over gene expression. Many viruses possess means to precisely regulate the rate at which the cell manufactures virus-specific products, the order in which gene products are expressed, and also provide some of their own enzymes necessary to replication. The extent and means of regulating gene expression are as diverse as the kinds of virus, and each has developed to meet the needs of a particular virus within the constraints imposed by the host cell. For it must be borne in mind that eukaryotic cells have not remained totally passive victims to the self-serving strategies of virus replication. Politicians do not provide a free lunch; neither does nature. In many instances it seems likely that certain viruses have been made to be of some use to the host. For others, the cell possesses a number of retaliatory mechanisms for minimizing the deleterious effects of virus infections. In many instances the cell has in effect learned to accommodate, or at least to tolerate, the presence of virus. Finally, multicellular organisms have perfected an organ system specifically for the purpose of recognizing and destroying harmful microorganisms.

This chapter will discuss the state of our current understanding of what viruses are, how they replicate, and how they affect their host cell. More importantly, an attempt will be made to relate the molecular biology of the virus replication process to the impact that such events have on the whole organism—the pathogenesis of virus diseases. For the most part the discussion will concentrate on viruses of eukaryotic cells, with an especial effort to consider animal viruses. Other virus systems will be referred to only when they offer a particularly interesting example of the subject under consideration.

VIRUS STRUCTURE AND FUNCTION

The essential feature of the intracellular phase of most viruses is that the nucleic acid core is constructed and organized to allow full expression of the genetic information. In some instances it appears that certain viruses function intracellularly as isolated nucleic acid free of other virus components. Usually, however, other proteins remain associated with the genome for purposes of protecting the nucleic acid from digestion by cellular enzymes, or because the proteins are necessary to appropriate expression of the genetic message.

The extracellular phase of a virus (the virion), on the other hand, reflects a much more complicated structure. Elements necessary to protect the genome from the various hazards encountered in extracellular fluids and in the outside world are ordinarily required. An exception are the viroids, thought to be small pieces of naked RNA but which are capable of infecting, then multiplying and damaging certain kinds of plants (potato spindle tuber viroid, chrysanthemum stunt viroid, and other exotica). No such examples are known for animal hosts, however.

The simplest virions are those consisting of a protein shell tightly surrounding the nucleic acid core. A number of features can be quickly inferred from such a structure. First, the shell or coat must be complete and tightly constructed. As the structural materials for the shell are virus specified, only a few species of proteins at most are available, so that, second, the subunit structures must be regularly repeating entities (capsids) held together by intermolecular forces. Third, this endows upon the structure the quality of symmetry, and much has been made of the kinds of the icosahedral or helical symmetries possessed by such virions. Fourth, assembly of the virion must be a relatively simple process rather than requiring elaborate schemes involving complex regulation; the simplest of these is "self-assembly," in which virus components naturally come together as a compact entity. Otherwise, it presumably would require more energy to keep dispersed molecules intrinsically destined to interact. Fifth, such self-associating structures must be compactly constructed; as such, many are extraordinarily stable to such forces as chemicals and heat. Simple shelled structures are characteristic of viruses which infect via the gastrointestinal tract. Presumably, the stability endowed by this form enables them to survive passage through the acid environment of the stomach.

The need to form compact and stable structures presumably limits the feasible size attainable in terms of dimension and genetic information. The adenoviruses are large enough that their genome consists of 30 to 40 genes, but other of these viruses are limited to approximately 10 genes. Finally, these viruses encounter a barrier in terms of release from the host cell. It appears the only way to get from the intracellular site of assembly to the extracellular space is by disrupting the cell membrane. In general this is accomplished by destruction of the membrane, brought about by severely impairing cellular metabolism and causing cell death. As a consequence such viruses generally tend to be cytocidal.

Perhaps as a means to avoid this outcome, and in recognition of the fact that it is undesirable for the parasite to kill an accommodating host, many viruses use a mechanism of *envelopment* as a means of egress from the host cell. This process adds an additional element to the final structure of the virion, an element, moreover, which as a rule is critical to infectivity of the virion. The reason for the latter will become obvious when virus-cell interactions are discussed; the formation of the envelope will be considered in detail in the next section.

Appropriately, viruses with envelopes are referred to as "enveloped" viruses. Those without are designated "naked" viruses. A desire to avoid destroying the cell is actually not the sole reason for envelopment, apparently, as a third group of viruses, the complex animal viruses known as poxviruses, can independently direct the manufacture of their own envelopes. Be that as it may, the envelope thus becomes the surface barrier presented to the environment. This appears to relax the critical relationship of internal structures to each other so that additional elements may be present in enveloped virions. Arenaviruses, for example, incorporate host cell ribosomes, and cellular genetic elements are frequently included in other viruses. Many virions may contain more than one genome copy (multiploidy). The relationship of protein to nucleic acid is usually not as rigid as with the naked viruses, so that the protein in many instances does not completely enclose the nucleic acid. The envelope contains lipid and protein in an arrangement characteristic of any cellular membranous structure, and is not nearly as resistant to destruction as are the icosahedral shells of naked viruses. Enveloped viruses are thus forced to be more cautious in their environmental exposures, and tend to prefer transit from host to host by direct contact. Spread via respiratory tract secretions is apparently also tolerable, as many of the viruses transmitted in this manner are enveloped.

The nucleic acid genome of viruses may be composed of DNA or RNA. All families of DNA viruses but one possess DNA as a double-stranded (ds) molecule; the exception is parvovirus, which contains single-stranded(ss) DNA (hepatitis B virus seems to possess dsDNA containing a long stretch of ssDNA). Conversely, all RNA virus families possess ssRNA genomes except the reoviruses, which carry their genetic information in the form of dsRNA. As far as is known all DNAs are one continuous molecule. Presumably this reflects the fact that eukaryote cells are familiar with this form of DNA and routinely are able to select only those portions needed at the moment. This is, of course, not the case with RNA; cell experience is limited to interpreting discrete individual messages on ss molecules. Reflecting this fact, many of the genomes of RNA viruses are segmented, each segment being monocistronic in the same manner as the cell's mRNA. Other RNA viruses, however, do not contain segmented RNA, but the molecule is continuous and polycistronic.

For the segmented RNA viruses, the virion RNA may be directly translatable to virus polypeptides, or a complementary copy (cRNA) must first be synthesized and the cRNA is translated to authentic virus polypeptides. This feature necessitates the assignment of polarity to RNA genomes; those in which virus RNA is directly used as messenger are said to be of messenger or plus-strand polarity. Viruses containing RNA which must be transcribed to cRNA to form functional message are, by convention, designated as negative-strand viruses and the RNA is said to be of antimessage polarity. The segmented ssRNA viruses are all negative stranded, with the exception of Nodamura virus (a virus containing two segments of (+) strand RNA, which along with a black beetle virus probably represents a unique group of RNA viruses). Nonsegmented ssRNA viruses may be plus or negative stranded. All known dsRNA viruses contain segmented genomes, almost certainly because nature is otherwise very naive in being able to interpret dsRNA information.

For all viruses the nucleic acid has a very specific configuration in the virion, usually with associated proteins which may have various functions. The DNA in viruses such as adenovirus, papovavirus, and herpesvirus are maintained in nucleosome-like structures, the DNA being wrapped around either virus-specified polypeptides or histones requisitioned from the host cell. Numerous other proteins are also associated with the nucleic acid core of viruses, which may function as enzymes, structural components, or regulatory molecules.

Naked viruses complete their structure, as previously mentioned, by aggregating a shell (or shells, in the case of reovirus), around the core (Fig. 1). The completed shell is the *capsid*, subunits of the shell are called *capsomeres*, and the structural polypeptides thereof are *capsid* (or coat) *proteins*. More than one species of capsid protein may be present in some viruses, but as a rule economy is effected by most in that only one or a very few different proteins compose the capsid. Further, many capsid proteins serve additional functions other than structural. As they form the exterior surface of the virion certain capsid proteins are of vital importance to virus-host cell interactions and to the early events during infection. However important a stable and nonyielding capsid is to the virion, it must also be readily removable once infection has been initiated.

Enveloped viruses (Fig. 2) generally do not have as rigid a nucleocapsid structure since the external protective structure is the envelope. The latter is derived from the membrane of the infected cell and consists of the lipid bilayer synthesized by the cell plus polypeptides which have, for the most part, been specified by the viral genome. Envelope polypeptides are of two classes: surface proteins which are almost always glycosylated by host cell enzymes with carbohydrate moieties, and an underlying protein layer called M (matrix or membrane) protein

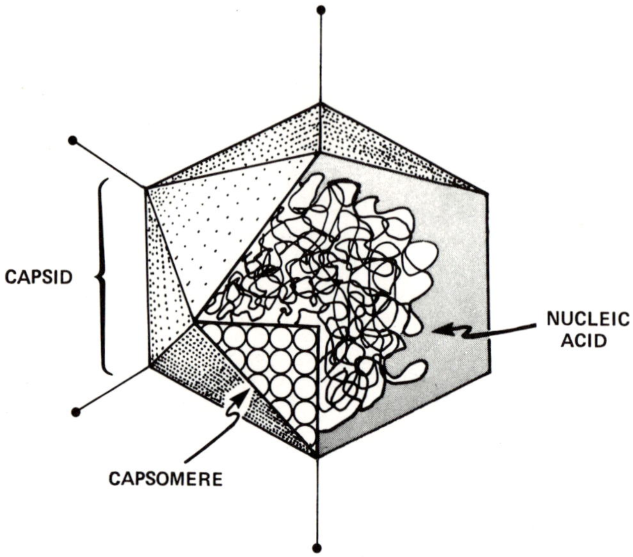

Figure 1 A naked virus, here illustrated with icosahedral symmetry.

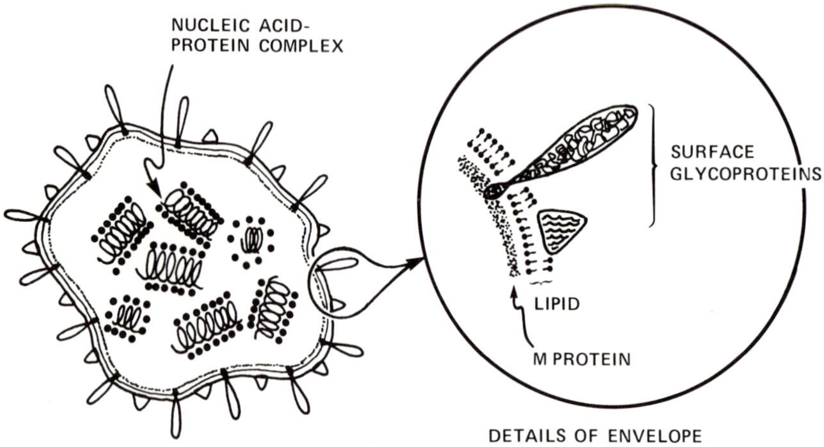

Figure 2 Diagram of an enveloped virus. Magnified section illustrates detail of envelope structure (see text).

Table 1 Virus Families: Features and Common Examples

Family	Genus	Genome	Envelope	Examples
Adenovirus		dsDNA		Human adenoviruses types 1-34; canine hepatitis virus
Arenavirus		ssRNA, segmented, -	+	Lymphocytic choriomeningitis, Bolivian hemorrhagic fever, Lassa
Astrovirus		?		? Human diarrhea viruses
Bunyavirus		ssRNA, segmented, -	+	California encephalitis, LaCrosse, snowshoe hare
Calicivirus		ssRNA, +		? Human diarrhea viruses (e.g., Norwalk, Montgomery County, etc.), vesicular exanthem of swine
Coronavirus		ssRNA, +	+	Human respiratory and diarrhea (?) viruses, mouse hepatitis
Herpesvirus		dsDNA	+	Herpes simplex, varicella zoster, cytomegalovirus, Epstein-Barr virus; pseudorabies, Marek's disease virus
Orthomyxovirus		ssRNA, segmented, -	+	Influenza viruses of humans, swine, ducks, horses
Papovavirus		dsDNA		BK, JC viruses of humans; SV40 polyoma virus
Paramyxovirus		ssRNA, -	+	Parainfluenza viruses, mumps, measles
Parvovirus		ssDNA		Adeno-associated virus

(continued)

(Table 1 continued)

Family	Genus	Genome	Envelope	Examples
Picornavirus	*Rhinovirus*	ssRNA, +		Human rhinoviruses, foot-and-mouth disease virus
	Enterovirus			Poliovirus, coxackie A and B virus, echovirus, enterovirus types 68-71,? human hepatitis A
Poxvirus		dsDNA	+	Smallpox, monkey pox, vaccinia, orf, rabbit myxoma, canary pox
Reovirus	*Reovirus*	dsRNA, segmented		Human reoviruses
	Orbivirus			Bluetongue virus, African horsesickness virus
	Rotavirus			Human diarrhea viruses
Retrovirus		ssRNA, +	+	Visna, maedi, Zwoegerziekte; mammalian, avian tumor viruses
Rhabdovirus		ssRNA, -	+	Rabies, vesicular stomatitis virus
Togavirus	*Alphavirus*	ssRNA, +	+	Eastern, Western, Venezuelan encephalitis viruses; Sindbis virus, Chikungunya virus
	Flavivirus	ssRNA, +		St. Louis, Japanese B encephalitis viruses; dengue virus, yellow fever virus

Unknown: hepatitis viruses (human B, ground squirrel, woodchuck); Marburg-Ebola viruses; Nodamura and black beetle viruses.

(not present in coronaviruses and togaviruses). Surface glycoproteins are attached, via strongly hydrophobic portions, into the lipid membrane or are inserted through the lipid bilayer and anchored to the M protein. Envelope glycoproteins are glycosylated as a consequence of the manner in which they are synthesized (see the next section). They also play critical roles in the infectivity of the virus.

In addition to these two basic schemes, other more elaborate forms are found in nature. The poxviruses are the largest of the animal viruses; their genome of some 240 genes ensures considerable control over their replication, and a complex structure.

To facilitate further discussion and to provide a perspective for the reader, Table 1 presents a current list of virus families, with some details of the features of each. In some instances popular examples are representative of genera so that these families are accordingly subdivided. The classification of viruses is as uncertain as for the rest of biology, and the table does not claim authority over convenience.

THE INFECTIOUS CYCLE

The virion represents the vector mode of transmitting functional genetic information from cell to cell and host to host. The host cell, on the other hand, can be conceptually visualized as a huge self-contained warehouse of metabolic activity, responsive to the regulatory mechanisms of the organism but capable of autonomous function. It is access to this vast treasure of biochemical wares which the virus seeks. Extracellularly, the viability of the inert virion is measured in finite terms, and destruction awaits only any one of a variety of inimical environmental hazards. Intracellularly, in contrast, the virus is a functional entity reproducing its own kind or secreted away for future appointments.

The effect which a given virus will have on the host cell depends on a number of factors. First, not all viruses are able to replicate in all cells. On the contrary, most viruses have very strict specificities in terms of the species of host which they normally infect, and the cell type in that host which is able to support replication. The inability to replicate in a given cell may rest with physical attributes of the virus or with biochemical characteristics of the cell. The overall concept is known as "permissiveness," indicating that a given virus can successfully initiate and carry out replication to completion in such a cell. Since viruses require so much assistance from the cell, nonpermissive conditions to any one of the replication steps will in general completely block the production of progeny.

In terms of infection of the host, the most critical step is the ability of the virus to seek the appropriate pathway from the route of entry to permissive cells. In almost all instances this simply involves tissue sites immediately adjacent to the portal of entry. In the respiratory tract at least different anatomic levels may be involved. Rhinoviruses infect and replicate best in the epithelial cells of the nasal mucosa; most other respiratory viruses must traverse the upper airway and initiate infection in the nasopharynx. While the mucous blanket may create an apparent barrier it must function primarily to trap larger particulate matter in the inspired air; viruses seem readily able to penetrate this layer and reach underlying tissue.

Thus, while the journey from one host to another may be arduous, viruses seldom have difficulty in reaching susceptible cells once they have attained access to the host. The next step enlists a recognition mechanism by which the virus is able to discern cells capable of supporting its replication.

Adsorption

The presence of molecules on the cell surface which can specifically interact with the virion surface is a major determinant of tissue tropisms by viruses. Without such an interaction the virus cannot adsorb to the cell surface, and without adsorption infection of the cell is not initiated. It might be noted that experimentally this requirement can be circumvented for some viruses in that the isolated virus nucleic acid may be infectious in cells otherwise nonpermissive (because they lack appropriate receptors). There is, however, little if any evidence that this occurs in nature (although it remains an intriguing possibility).

The identity of the receptor molecules on cell surfaces are poorly characterized, but some information is available. From what has already been discussed it can be inferred that in the absence of receptors the cell is not susceptible to infection by that virus. Different viruses, however, use different receptors (although not exclusively so) so that, on this basis, a particular cell may be susceptible to infection by one virus and not another. Relatively speaking, virus receptors seem to be plentiful and are randomly dispersed over the cell surface. It is estimated that susceptible cells possess on the order of 10^4 receptors per cell for picornaviruses, while the number of parvovirus receptors on mouse cells has been determined to be approximately 10^5 per cell.

The receptor molecule on the virion is prominently displayed on the surface and is designed to interact specifically with its counterpart on the cell surface. This affixes the virion and apposes virion surface to cell membrane as a prerequisite to the molecular interactions leading to entry. In actual fact, adsorption is the initial step leading to all events of the replication cycle, but the process occurs as a continuum of activities rather than as a series of discrete processes. The behavior of picornaviruses on cell surfaces serves as an example.

Most coxsackieviruses reaching the surface of a susceptible cell remain associated with the membrane. Those that bind to receptor sites immediately lose one of the capsid proteins (VP-4), converting them to so-called A particles. A particles may then be interiorized and begin replication, but the majority elute from the cell surface into the extracellular environment, where they remain as now noninfectious particles.

There is much evidence to point to the fact that the presence of cell surface receptors are major determinants of susceptibility. Even though for some the virion passively gains access to the cytoplasm of the cell by virtue of cellular endocytosis of the particle, it is likely that adsorption is a prerequisite. Presumably, molecular interactions significantly perturb membrane fluidity and organization and trigger active cell responses much as are seen with capping phenomena in lymphocytes, for example. The profound effect by even a few adsorbed virus particles is illustrated by EMC virus (a picornavirus); adsorption of as few as five virions per cell results in a measurable increase in membrane fluidity secondary to protein rearrangement.

In the example given above it was seen that picornavirus adsorption occurs concomitantly with structural alterations of the virus, lending a further aspect to the specificity of initial interactions. Thus, not only the appropriate spatial configuration (to effect covalent adsorption) but the appropriate biochemical activity (to remove one of the coat proteins, probably by enzymatic action) seems to be required. Even so, it further seems likely that many cells possess the proper receptors necessary to the initiation of infection yet are deficient in the requisite metabolic factors which would ensure complete replication intracellularly. Whether natural selection processes would allow abortive infections such as this to be perpetuated is problematic. In fact, in some situations passive adsorption alone may be harmful to the host. Newcastle disease virus will adsorb to splenocytes of mice. Eaton has demonstrated that this alters the antigenic recognition of these cells sufficiently that mice develop immune reactivity to normal spleen cells which may progress to an autoimmune process and death.

Otherwise, in examples of successful adsorption studied thus far, those which turn out to be to otherwise nonpermissive cells seem nevertheless to offer advantages to the virus natural history in other ways. Thus, the ability of influenza A virus to adsorb to group B streptococci may offer a means of transporting the virus to weakened host tissues, or alternatively, simply represent a temporary haven. Similarly, even though influenza virus does not ordinarily replicate in macrophages, the peripatetic nature of the latter may serve as a means of disseminating adsorbed virus to distant tissue sites. Actually, while this example may be true for adults (humans or mice), macrophages from newborns may serve double duty for viruses such as herpes simplex virus, in that the macrophages not only possess receptors but also are permissive. Here, then, the macrophage both supports the replication and disseminates the virus. Curiously, with maturation (2 to 3 weeks of age for mice) macrophages become nonpermissive.

In many instances the pathogenesis of a virus disease is clearly related to the tissue tropism of that virus, and the tropism has been found to be related solely to the presence of virus receptors on the surface of component cells. Recently it has been found that infection of mice with encephalomyocarditis or coxsackie B4 virus results in diabetes mellitus in certain strains of mice, but that susceptibility to developing diabetes is genetically controlled. In turn, it is now known that susceptibility is controlled by a single autosomal gene and is phenotypically expressed by the presence of receptors on β cells of the endocrine pancreas. Thus the expression of the susceptibility gene product places virus receptors on the cell surface, allowing virus to adsorb, infect, and destroy that specific cell. In the absence of β cells, of course, no insulin is produced and diabetes develops.

In similar fashion, the distribution of receptors for herpes simplex virus in the nervous system may be a major determinant of the pathogenesis of these infections. Receptors are more plentiful on synaptic portions of sensory nerve cells (allowing access to the intra-axonal compartment from whence they are transported to their site of latency) but are absent on the perikaryon (preventing cell-cell infection in the ganglion itself and thus preserving latency). In the central nervous system, on the other hand, both neuronal and glial cells possess receptors, presumably rendering both susceptible to infection and accounting for the particularly necrotic nature of herpes infections of the central nervous system.

Similar examples are known for a number of other viruses. Receptors for reovirus type 1 are found limited to ependymal cells lining the ventricular system of newborn mice, whereas receptors for type 3 are found throughout the central nervous system. This instance also demonstrates the reciprocal consideration that receptor molecules on the virion surface themselves obviously possess specificities. The surface molecule of reovirus designated HA (hemagglutinin) is the virion receptor. HA of reovirus 1 can adsorb to ependymal cells only and produces inflammatory disease of that structure. The HA of reovirus type 3 is more versatile; owing to its ability to adsorb to a variety of cell types in the central nervous system, infection results in diffuse encephalitis.

Virion receptors of enveloped viruses are, not surprisingly, glycoproteins, a fact established for herpesviruses, alphaviruses, flaviviruses, rhabdoviruses, myxoviruses, paramyxoviruses, and retroviruses. Although it is unknown whether it is of natural significance, experimentally it can be shown that the assumption that virion receptors are essential to infectivity is true. Such data, moreover, give some insight into how adsorption promotes infection with enveloped viruses, although there is conflicting evidence for certain viruses. The envelope glycoprotein VP-7 of herpes simplex virus is required for infectivity but seems to function by promoting fusion between the envelope of the virus and the cell membrane. Further, glycoprotein gE of the envelope is an Fc receptor. While various physiologic roles have been proposed for such a molecule, one suggestion is that gE mediates attachment via β_2-microglobulins which are found on most cellular surfaces, and possess domains similar if not identical to immunoglobulin Fc regions. As such gE would thus only facilitate primary apposition to the cell surface. The primary determinants of host range of herpes simplex would still depend on (1) appropriate virus receptors on the host cell, (2) appropriate receptors in the envelope, and (3) the presence of the virion envelope glycoprotein able to promote fusion with the host cell. Thus, virus factors as determinants of pathogenesis based on virion surface components may be more complex than simply the presence of absence of the specific receptor.

Just as was previously stated, that certain virus families seem to acquire an envelope in part to solve the problem of harmlessly egressing from the cell, some guarantee must exist that the envelope does not pose an insurmountable problem to the successful reentry of viral nucleic acid back into fresh host cells. It does not. In fact, for many enveloped viruses, entry into the cell may be the precise mirror image of exit from the cell. These viruses enter by fusion of the virus envelope with the cell membrane (Fig. 3C). Initial fixation is no doubt accomplished via receptor-receptor interactions. Eventually the virus envelope (with its viral glycoproteins) becomes incorporated into the cell membrane. The nucleic acid thus finds itself free on the cytoplasmic side of the membrane. The molecular basis of this process is discussed in more detail in this section, Viruses and Cell Membranes. A major difference, however, is that some stimulus must exist which initiates intermolecular mixing of the envelope and the cell membrane in the first place. While the process may be not understood, fusion of lipid bilayers is a common enough process in the daily activities of eukaryotic cells (phagocytosis, endocytosis, and cell division, for example). Thus, appropriate stimuli are probably readily available to enveloped viruses which have been adapted to this method of entry. It may be that perturbations induced by receptor attachment result in interiorization of receptor complexes, pulling in the envelope along with the glycoprotein molecule and allowing membrane-envelope mixing (Fig. 3C). At least for a time, envelope glycoproteins can be demonstrated on the cell surface after the nucleic acid core has entered the cytoplasm; while this is only temporary, the cell is susceptible to immune attack while these virus antigens remain displayed on its surface.

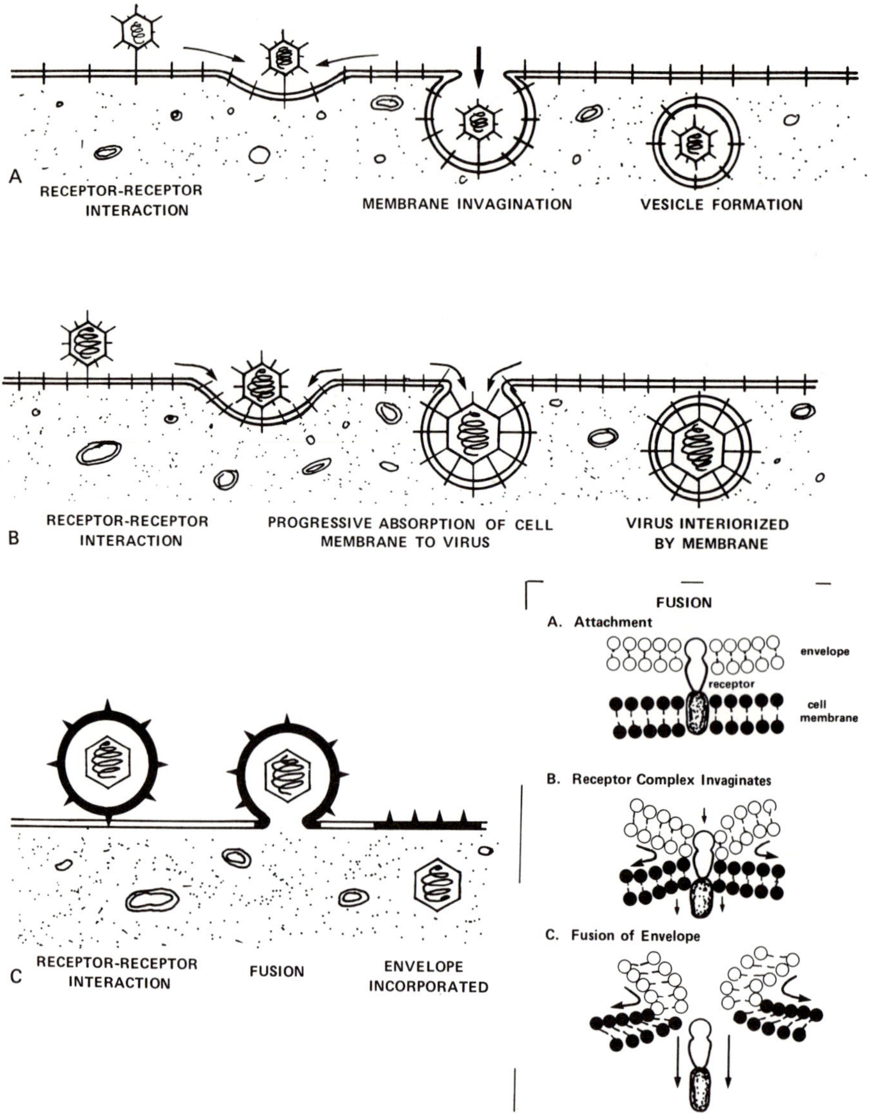

Figure 3 Diagrammatic details of virus adsorption. (A) Endocytosis. (B) Viropexis. (C) Fusion. Inset diagrams a possible sequence of membrane-envelope interactions which might occur with fusion.

It does appear, however, that not all enveloped viruses enter by this fusion process. Some appear to be interiorized by other means. Less than 3%, if any, of VSV and Sindbis virions enter cells by fusion. Alternative means of entry include enzymatic dissolution of the envelope at the cell surface and active ingestion of the core, and endocytosis of the whole virion with degradation of the envelope within the cytoplasmic vacuole thus formed. Obviously, both processes are essentially the same. However, endocytosis of intact virions has been frequently visualized, whereas surface deenvelopment has not. Explanations for failure to find the latter include the possibility that endocytosis activates appropriate enzyme systems, or that the ingestion of whole enveloped particles is a long-lived stage which may or may not lead to replication. Nevertheless, current evidence finds a correlation between the ability of viruses to cause syncytia formation (see the next section) and entry by fusion. Presumably both effects are mediated by the same molecules, and for Sendai virus (a parainfluenza), for example, both activities have been defined for the F surface glycoprotein (see below). Finally, viropexis is a term used to denote a process of entry in which there is progressive receptor-receptor interaction until the virion is enveloped in cell membrane (Fig. 3B. Viropexis differs from endocytosis primarily in that it is a passive process, with receptors on the cell membrane sequentially and circumferentially interacting until the membrane surrounds the virus (Fig. 3B). In contrast, endocytosis is an active cell process, requires energy, and is mediated by contraction of microfilaments.

The function of the cell membrane molecules that otherwise serve as virus receptors is largely unknown. Almost certainly these molecules exist primarily to serve a vital function to the cell. Otherwise, the disadvantage bestowed upon the host by such a determinant of susceptibility to harmful infection would be rapidly lost. For some viruses receptors are found on numerous cell types. Such is the case with paramyxoviruses, which use neuraminic acid-containing glycoproteins and glycolipids; here there are additional determinants to susceptibility, as we shall see later. It has been reported that the alphavirus Semliki Forest virus uses as its receptor products of the major histocompatibility complex on mouse cells and the counterpart HLA system on human cells. Such an hypothesis is attractive inasmuch as this would provide a ready supply of susceptible cells to this virus and correlates nicely with the broad range of cells susceptible in vitro to infection with the virus. Unfortunately, more recent data dispute this hypothesis. Other viruses may be highly specialized owing to the narrow specificity of their cell receptors. Infectious bursal disease virus of chickens (a reovirus) appears to use immunoglobulin M as its receptor. As a result, it infects only B lymphocytes bearing surface IgM molecules (and eventually leads to lymphocyte necrosis, bursal atrophy, and immunosuppression).

The significance which the nature and location of cell receptors carry for the pathogenesis of virus disease is perhaps best illustrated by an hypothesis proposed by Holmes and his colleagues for rotaviruses. They suggest that lactase present in the brush border of intestinal epithelial cells serves as the receptor for these viruses, which are major causes of diarrheal disease in young infants and certain animal species. Only in these cells is the enzyme present as a surface membrane component. Lactase as the receptor for rotaviruses would account for such features as localized pathology (distal villi of the small intestine), age susceptibility (intestinal lactase is lost in older animals and members of some human ethnic groups), and treatment considerations (affected individuals are temporarily lactase deficient as the result of their infection). Obviously, intestinal lactase must be present in neonates to digest the major disaccharide in maternal milk. It is interesting to speculate whether the loss of lactase with age represents the result of selective pressure against susceptibility to infection in these individuals.

Intracellular Replication

Uncoating

A major step has been accomplished. The virion has successfully reached the site of permissive cells in the host, has recognized those cells, and has been able to physically associate with the definitive elements so critical to their continued life history. Before replication can begin in earnest the final vestiges of the virion's extracellular sojourn must be shed. For the most part this involves removal of coat proteins. This is usually accomplished concomitantly with transport to the definitive site of replication. At the same time, uncoating not only exposes the nucleic acid as a biologically functioning molecule but frequently also activates virion-associated enzymes required to initiate replication.

Once the interior of the cell has been reached, the primary consideration of the virus becomes the appropriate site at which replication will occur, and transport to that region. For most DNA viruses this means the nucleus, with the poxviruses being a major exception. Conversely, RNA viruses (except perhaps the orthomyxoviruses) will replicate in the cytoplasm, which no doubt simplifies matters for them. Accordingly, these will be considered first.

Obviously, once past the cell membrane, the cytoplasmic viruses are immediately at potential replication sites. Moreover, for the picornaviruses the evidence is good that most of the coat proteins have been removed by membrane-associated enzyme systems. This releases genome RNA directly to the cytoplasm where, once associated with membrane-bound ribosomes (rough endoplasmic reticulum), translation to gene products can begin directly. For other RNA viruses the situation may not be so straightforward. This in large part depends on the genome organization and will be considered in detail below. Nevertheless, uncoating of RNA viruses in general can be considered as a membrane or cytoplasmic event, uses host cell enzymes already present at the time of infection, and is not a major event in considering pathogenesis of virus disease. It should be noted, however, that the uncoating event may be susceptible to chemotherapeutic attack, as witness the efficacy of the drug amantadine in inhibiting the uncoating of certain RNA viruses.

A major exception to the relatively simple and nonspecific nature of uncoating and entry of RNA viruses is found with the myxo- and paramyxoviruses. It will be recalled that these agents adsorb to glycosylated proteins containing a terminal N-acetyl neuraminic acid. As these moieties are found on the surface of virtually all vertebrate cells, yet the number of cell types which support replication of paramyxoviruses is limited, there are obviously additional determinants of infection. One of these is the presence on the cell surface of a protease which is capable of cleaving the F_0 glycoprotein located on the virion surface. This molecule does not participate in adsorption, but cleavage results in an activated product responsible for initiating envelope fusion with the cell membrane. This step is essential to infectivity; in the absence of an appropriate cellular protease F_0 is not activated and infection will not occur. Once activated, however, fusion ensues and the viral genome gains access to the cell interior. This process, then, defines the host range of many of the paramyxoviruses, as well as tissue tropism and virulence.

Viruses destined to deliver their genomes to the nucleus face a special problem. Not only must they traverse the surface membrane of the cell, but they must also gain access to the nucleus by crossing the nuclear membrane. The latter is a rather complicated structure composed of an outer and inner nuclear membrane, separated by the intracisternal space and punctuated by nuclear pores. Rather than being avenues for entry the pores instead are sufficiently small as to prevent diffusion of molecules greater than 70,000 daltons. An approach similar to that used at the surface membrane is employed by various viruses to breach this nuclear barrier. Adenoviruses, for example, gain the cytoplasm qualitatively intact (probably by viropexis), where they are then transported to the nuclear membrane. This occurs rapidly in association with microtubules and leads virions to the nuclear pore complexes. Uncoating occurs at or near this region, and the DNA genome is delivered through the pore into the nucleus.

The papovavirus SV40, on the other hand, employs cellular membranes to move from the extracellular environment to the interior of the nucleus by a series of fusion and fission steps. Although extracellularly a naked virus, the cytoplasmic stage of entry finds SV40 tightly enveloped by invaginated cell membrane. The membranous virion is transported through the cytoplasm and apposes to the membrane of the rough endoplasmic reticulum or the outer nuclear membrane. This releases the naked virion into the intracisternal space, and from here the virus readsorbs to the inner nuclear membrane, again presumably via virus receptors on the latter structure. This event must in some way induce disorganization of the dense lamina and heterochromatin which line the membrane, and allow it to invaginate. Once in the nucleus biochemical replication ensues. This argues for an enzyme system present in the nucleus, but not the cytoplasm, which can uncoat membranous virions, a factor of course fortuitous for the virus as this is precisely where replication must occur. It should be noted, however, that Maul and his colleagues have noted SV40 within mitochondria, another environment which fosters DNA replication.

The movement of viruses to favored compartments within the cell has been briefly mentioned. In some instances, as already discussed for picornaviruses, the point of entry and the replication site may be contiguous. In other instances simple diffusion may accomplish the necessary movement. Of prime importance, however, appears to be the microtubule component of the cell cytoskeleton. A number of viruses have specialized capsid proteins which function to bind the virion to the microtubule. Adenovirions attach via the hexon capsomere to a high molecular weight protein associated with the microtubule, and in this way are transported from the cell membrane to the nucleus. Reoviruses also bind to microtubules; this is mediated specifically by a minor outer capsid polypeptide σ_1 (which also functions as the cell receptor). In this instance, however, it appears that type 1 σ_1 (reovirus serotype 1) binds much more readily than reovirus type 3 σ_1. Furthermore, these dsRNA viruses replicate in the cytoplasm. The fact that microtubules associate with the replication sites of type 1 but not type 3 suggests that this represents a facultative relationship to virus multiplication. Nevertheless, the fact that microtubules are instrumental in the movement of cells and their internal components suggests that virion binding in many instances serves as the mechanism of internal transport.

Uncoating of the virion is obviously a prerequisite of active participation of the virus itself to the replication process. As such it requires that the enzymes or factors necessary to the uncoating process be contributed, presumably passively, by the cell. The ability of the cell to participate is another expression of the permissive state. It is also a process which must be suppressed later on in the replication cycle if progeny virus are to be successfully assembled and released. This illustrates the need for the virus not only to exploit the macromolecular processes of the cell but also to exert some degree of control over its host. Accordingly, the virus must carefully intrude into the cellular environment so as not to seriously upset its metabolic processes but at the same time subvert them to its own needs.

Genome Expression

The pathogenesis of virus infections ultimately bears on how the virus affects cellular metabolism, why it exerts this effect, and which cells are thus affected. To understand each of these questions has required detailed knowledge of both cellular metabolism and virus replication. While this has not yet been fully attained, sufficient information is available to appreciate many of the features of each and how they interrelate. A brief summary will be presented here so that the foundations of the numerous concepts encompassed in virology can be better understood and, it is hoped, the majestic systems which have evolved during the history of virus-cell interactions will become apparent.

ssDNA Viruses. The ssDNA viruses are chosen to begin the discussion of replication because they are among the simplest of all viruses and, as currently understood, are represented by a single genus (parvovirus). They are small naked viruses the DNA of which is encapsidated as a single strand of DNA. Examples in nature are the minute virus of mice (MVM), hamster H-1 virus, adeno-associated viruses (AAV), Kilham rat virus, and Aleutian disease virus (ADV) of mink. Two major classes exist; those which are able to replicate completely by their own means in a permissive cell (the "autonomous" parvoviruses) and the *defective* parvoviruses which require the assistance of *helper* viruses. All replicate in the nucleus, specifically because they rely on the host cell's DNA polymerase to synthesize progeny virus DNA and RNA polymerases to manufacture viral mRNA.

ADV and MVM virions contain only noncomplementary strands of DNA; that is, each is of the same polarity and, in fact, appears to be the minus strand from which mRNA is to be transcribed. This contrasts with adeno-associated virus, in which virions package only one DNA strand but they are approximately equally divided between plus and minus strands. Extraction of the DNA from virions thus yields dsDNA. (Such multipartite genomes are not uncommon; certain examples are known to exist for plant viruses in which separate genes are individually packaged in virions.)

The parvoviruses represent the simplest of parasites. AAV, for example, may be capable of coding for only one or two gene products. The polypeptides encoded by the parvovirus genome may thus be expected to serve more than one function, and there is good evidence that for some the coat proteins also are required for DNA synthesis. Further, the configuration of the DNA molecule itself has been economically constructed, so that although the ssDNA is a linear molecule, at the 3'-end the first hundred or so nucleotides are base-paired. This forms "hair pins," or looped-back regions, which thus become dsDNA in configuration. This probably serves as a primer to initiate complementary strand synthesis.

The replication of DNA is outlined in Figure 4. Two goals must be accomplished: synthesis of coat proteins and replication of minus strand DNA for progeny. Both are achieved by using cellular polymerases for the synthesis of the complementary strand of DNA. Actual synthesis may involve direct extension from the looped-back end, but for simplicity is illustrated as duplication of the entire minus (-) strand. The dsDNA intermediate is a familiar model to the cell, and it is presumably a simple matter for cell enzymes to transcribe and process RNA (messenger RNA) and complementary DNA (progeny genomes). Three or four species of viral RNAs are found in infected cells. They are transcribed from overlapping regions of viral DNA, either from a single promoter site or from splicing of common transcripts.

As the parvoviruses contain so little information in their diminutive genomes there is little control over the production of mRNA species either temporally or by quantity. Biochemical events rather appear to happen as they will; once sufficient raw materials are formed progeny virions are assembled.

AAV is defective in two steps, the initiation of DNA synthesis and a maturation function. Adenoviruses can provide assistance for both, so that coinfection of a cell with AAV and an adenovirus will allow replication of AAV. Infection with AAV alone thus represents insufficient genetic information for replication, as the virus is incapable of initiating DNA synthesis. (Herpes simplex virus can also provide helper functions in most cells but cannot assist maturation.) It would thus be problematic if AAV could survive in nature solely as a defective virus. However, as with most viruses, AAV has an operational contingency in the event that the helper function is missing. In some way the AAV genome stimulates a burst of cellular DNA synthesis. This in turn serves as an apparent prerequisite to incorporation of the virus DNA into the cellular chromosome, where it remains as an integral part. Virus DNA is thus replicated with cell division and awaits some future circumstance when it will excise from the cell genome and regain its status as an entity. This mechanism of *integration* is favored by many viruses as a means of *latency* or *persistence*, and preserves virus genetic integrity without the continual pressure of actively pursuing replication. The fact that virus and cell DNA can interact in this manner holds considerable significance as to the origin and function of viruses. It should be noted, however, that the autonomous parvoviruses do not appear to integrate.

The parvoviruses thus possess so little genetic information that they are extremely dependent on many cell factors. In general they

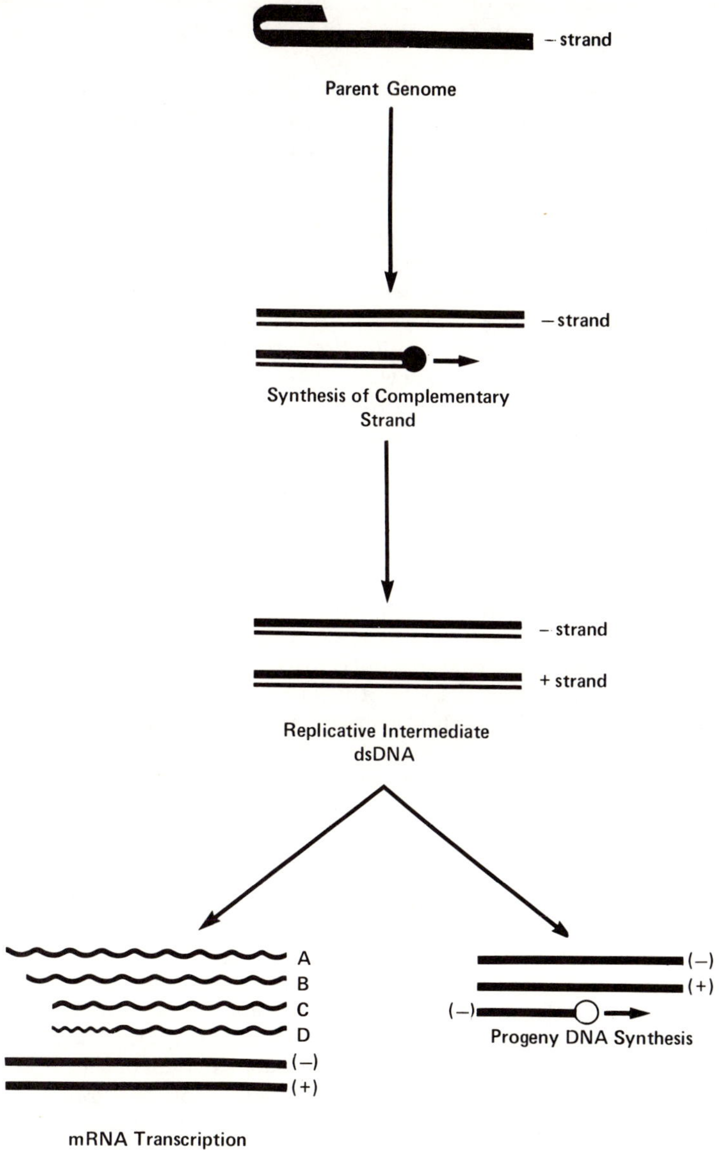

Figure 4 Replication of ssDNA viruses (see text).

are not major pathogens, perhaps because they would be so vulnerable to any resistance mechanisms of the host. ADV is, however, associated with a chronic disease in mink (see the next section IV).

dsDNA Viruses. This large group of animal viruses is represented by two families of naked viruses (adenoviruses and papovaviruses), one family of enveloped viruses (herpesviruses), and the complex poxviruses. They tend to be large in size and genome content, possibly because their genomes so closely resemble eukaryote genes for which the host cell already possesses a complicated system of information processing. Because of this similarity these viruses have served as models of gene expression. Consequently, a wealth of information exists as to how they replicate.

For convenience, intracellular replication steps are related to the temporal sequences preceding or following DNA synthesis. These are termed, respectively, early and late events. Genes expressed before DNA synthesis is initiated are thus called *early genes*, in contrast to *late genes*, which are expressed thereafter. In addition, some early genes are expressed before (or in the absence of) synthesis of virus proteins. This must be accomplished by cellular polymerases, which transcribe mRNA from virus DNA, and these genes are referred to as *immediate early* genes. In the immediate early phase of replication all events, regulatory or synthesizing, depend entirely on host cell factors. The products of these events are, of course, virus polypeptides, which then assume the role of controlling further gene expression and are involved in virus DNA synthesis. Late genes tend to be concerned, not surprisingly, with structural polypeptides and maturation of progeny virions. There is wisdom, then, to the arrangement, expression, and interrelationships of virus genomes. They rely heavily on the ability to manipulate cellular processes to their own purposes while ensuring that this subversion does not destroy the cell's capacity to support replication.

The dsDNA viruses possess luxuriant genomes compared to parvoviruses. The adenoviruses can code for 30 to 40 gene products, the herpesviruses 70 to 80, and the poxviruses approximately 200. In part this is used for a number of products which serve regulatory functions, but it also reflects more complex virions. Poxvirions, for example, contain approximately 100 polypeptides as structural units. The nucleic acid is arranged in all dsDNA viruses as nucleosome-like structures, with the virus genome core formed either on borrowed host cell histones or, at least for adenovirus, its own small basic proteins.

The DNA is not necessarily arranged in any stable pattern. For example, herpes simplex virus genome is represented by four genetically distinct molecules. The genome is physically divided by inverted duplications of the terminal segments into a long (L) and short (S) segment. The two termini of each segment possess the same sequence. As the DNA of each segment is oriented the genome can be made up of normally oriented L and S or permutations of each oriented in a normal or inverted manner. Functionally, however, a full complement of information is present in each isomeric arrangement. The gene arrangement and location, in general, finds various classes of genes (i.e., immediate-early, early, or late) scattered at separate sites on either strand of DNA. In other words, early genes are not clustered in one region but are dispersed throughout the entire genome. This is not invariable, however. Herpes simplex virus (HSV) genes for structural polypeptides which are to be topographically associated in the virion are found in clusters on the genome. Obviously, then, the native anatomy of the genome is not a major determinant of gene expression.

The entire course of gene expression involves a number of steps which closely parallel those found in normal cells. RNA polymerase transcribes a specific region of the viral genome to form RNA. This is processed extensively at the nuclear site to yield functional mRNA which is transported to cytoplasmic ribosomes. Here the mRNA is translated to a polypeptide product. Again, various modifications of the polypeptide may occur, and the polypeptide is transported back to the nucleus, where it may serve in regulating further biosynthetic events or as a structural component. Regulation of gene expression to the final product may thus be exerted at any of the several steps involved. Note, however, that poxviruses replicate entirely in cytoplasmic sites ("factories") so that macromolecular transport is not required.

The general scheme for regulation of the dsDNA viruses is the "cascade model" accepted for HSV. The products of immediate-early, early, and late genes are referred to as α, β, and γ proteins, respectively. The synthesis of α proteins turns on the expression of β, which in turn represses the production of α. The synthesis of β and γ polypeptides is similarly related, so that in the end the sequence α, β, and γ is expressed temporally. Synthesis is thus said to be coordinately regulated and sequentially ordered, and the peak production of each class of proteins thus appears in a well-defined episodic progression in the infected cell.

Transcription of immediate-early (IE) genes begins immediately after infection occurs, and is accomplished by an endogenous virion enzyme for poxviruses and by cellular RNA polymerase II for HSV-1. Specifically these genes and not others are expressed presumably because cell polymerases possess a greater affinity for the promoters of these sequences than for others. The primary transcript is actually a giant precursor mRNA which is considerably modified before it is ready to function at the ribosome (as with all eukaryote transcripts). Thus, mature mRNA is formed by "capping" the 5'-end with a methylated guanylic acid, methylating internal adenylic acid residues, and adding polyadenylic acid at the 3'-end. Processing is accomplished by cellular enzyme systems, except for the poxviruses, which again carry appropriate enzymes in the virion. Final nuclear modification involves splicing out unneeded internal sequences.

The early phase (pre-DNA synthesis) of transcription is turned on by IE products. With HSV, for example, it appears that an IE gene product (ICP4 or VP175, whose gene is actually diploid) is responsible for switching transcription from IE to early (E) genes. It does this either by binding directly to the promoter sequence of one set of genes or by complexing with RNA polymerase to alter its specificity. As a result, E genes are turned on and IE genes (including that for VP175) are turned off. Thus, IE gene products effectively serve as a prerequisite to the expression of E genes, but once IE proteins have been produced in sufficient quantity, cell energy devoted to the now unnecessary IE information is diverted to E.

Early mRNA of adenovirus is transcribed in four noncontiguous blocks of the genome; two are transcribed "left-to-right" on the *r* strand of DNA and two "right-to-left" on the *ℓ* strand. It appears, however, that the DNA sequences of both strands are transcribed; this yields RNA from complementary strands, which are, of course, themselves self-complementary. These *symmetric transcripts* have been found in cells infected with papovaviruses (polyoma and SV40), adenovirus, herpesvirus, and poxvirus. Functionally the symmetric transcripts are processed to asymmetric mRNA, but these dsRNA molecules may have considerable significance, as will be discussed in the next section.

E genes code for proteins which will primarily be involved in viral DNA replication, regulation of viral gene expression, and inhibition of cell DNA synthesis. The virus is concerned throughout this period in laying the groundwork for actual duplication of self through both virus DNA and protein synthesis. It must ensure that host cell processes will be properly modified in advance so that all the numerous steps will proceed smoothly. Just as importantly, however, the virus must anticipate potential problems. Thus, if conditions may not be proper for complete replication, for example, or even to the point that the host organism is mounting a sufficiently effective defense that the infection per se might be eradicated, alternate means must be available to preserve genetic information at the expense of shutting off further replication. For the most part these maneuvers can be classed as *persistence* mechanisms, and they will be discussed in Persistence. It is in the early stage of replication, nevertheless, that the virus must be prepared to switch from replication to persistence; thus, some E proteins participate in encouraging integration of the virus genome, are involved in *cell transformation*, or protect virus DNA from enzymatic attack by cell enzymes. Considerable coordination is also required, as a virus which must persist in the host cell for a prolonged period will not want to have seriously inhibited cellular synthesis before embarking on a path to latency.

The onset of virus DNA synthesis thus represents a monumental step in the replication cycle of dsDNA viruses. Early genes tend to be turned off (especially with herpesviruses); DNA for progeny viruses is being synthesized, thus expanding the gene pool of the virus; the host cell has been forced to pursue a course of action more to the benefit of the virus than to self; and late gene products are now committing the virus to the assembly of progeny. The stage is set for the individual cell to signal in whatever way possible to the organism that aliens are present and that the situation is potentially serious. The ability of the host to respond appropriately will probably determine the ultimate outcome of this infection.

DNA for progeny virions is replicated semiconservatively; that is, with synthesis of a new strand, for example, the old ℓ strand is displaced and used as template for new *r* strand synthesis. Thus, each new dsDNA molecule is composed of one original strand and one newly synthesized strand. Adenoviruses synthesize as an early gene product a DNA-binding protein which is present in approximately one molecule per seven nucleotides proportions and is involved in both initiation and elongation of replicating DNA (and may also protect virus DNA from digestion by cell enzymes). Virus DNA is replicated as rapidly as eukaryote DNA, approximately 2×10^6 daltons of new DNA per minute.

The signal for the initiation of DNA synthesis and switch to late genes is largely unknown. Presumably, early gene products play a role, but these have not been well defined. With adenoviruses most or all late transcripts are found initially as one giant mRNA (from 75% of the genome). This is then polyadenylated at the 3'-end, 1 of every 200 internal adenylates is methylated, and the molecule is capped. Then splicing occurs to form approximately 12 mature mRNAs, each with a common 5'-sequence, together representing only one-fifth of the original molecule. The viral mRNAs are transported to the nucleus, and the remainder of the initial transcript accumulates in the nucleus or is destroyed. Almost all late proteins are encoded on one strand of the genome, so that one signal thus acts to turn on synthesis of virtually the entire late class of genes.

Virus mRNAs are translated by cell ribosomes to virus polypeptides. These may then be modified by glycosylation, phosphorylation, or sulfation and transported back to the nucleus. Some of the late polypeptides may serve regulatory functions, especially phosphorylated proteins which may alter DNA binding affinity. For the most part, however, late virus proteins begin participating in the assembly of progeny virions.

ssRNA Viruses. Originally, viruses endowed with ssRNA molecules as the purveyor of genetic information most probably functioned directly as mRNA interjected into the cellular biosynthetic apparatus. At some time it was then discovered that cRNA copied from mRNA was also a feasible means of carrying information. Perhaps because it offered additional means of regulation or ease of expression this molecule was adapted to serving as the genome for many viruses. Viruses with RNA of positive polarity will be considered first.

Positive-strand ssRNA Viruses. These viruses are competing directly with cell mRNAs for ribosomes, and virus proteins are synthesized at the expense of cell proteins. As a result the + strand ssRNA viruses tend to have a profound effect on the health of their host cell while they are pursuing a course of replication.

Input (parental) virus RNA is translated directly to virus polypeptides. The RNA molecule is thus fully functional in the isolated state, and as such purified virus RNA is infectious (the only RNA viruses whose RNA is infectious). In addition to serving as messenger, the genome RNA must also serve as template for progeny virus genomes. Some sort of order must be imposed on these two functions so that they do not interfere with each other. This is initially guaranteed by the fact that RNA replication cannot be accomplished without a protein encoded by the RNA, so that only translation would be initially effected.

For picornaviruses, the general scheme for polypeptide synthesis is as follows (Fig. 5A). Virus mRNA is generally translated preferentially, in that it possesses a greater affinity for initiation factors. Ribosome binding occurs at a single site, even though, unlike cell RNAs, picornavirus RNA is not capped at the 5'-end. Translation proceeds virtually the entire length of the viral RNA, yielding a polyprotein which is the precursor for all virus polypeptides. The polyprotein is cleaved by a series of steps to form final polypeptides which function as structural, regulatory, and enzyme proteins for progeny synthesis. Cleavage schemes vary for different species of picornaviruses but generally follow a sequence of events which yield progressively smaller polypeptides.

Posttranslational processing is employed as a solution to the fundamental problem of how to express distinct messages on RNA molecules. The options are many, and in fact the various answers in themselves represent a study of the biology of RNA viruses. The difficulty with the polycistronic message per se seems to be that the eukaryote ribo-

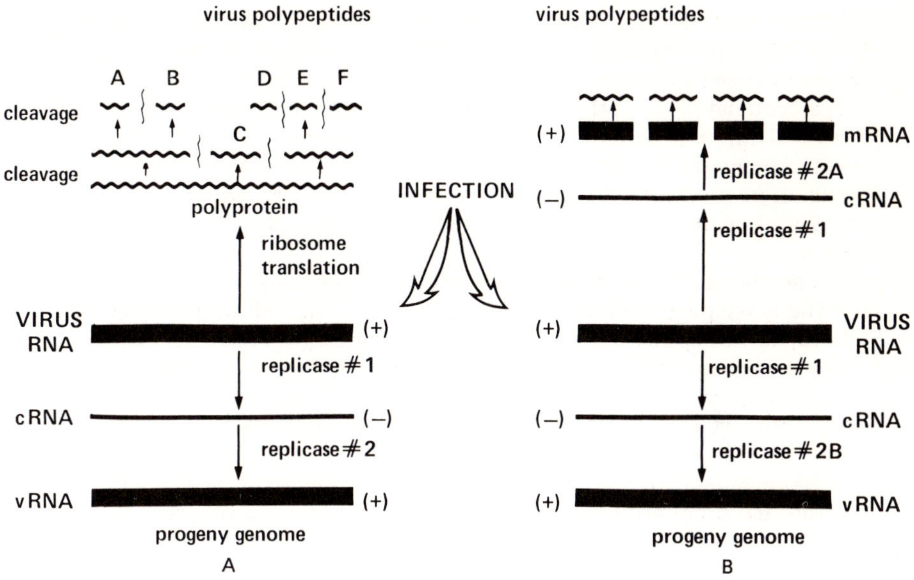

Figure 5 Schematic details of plus-strand ssRNA virus replication. (A) Picornavirus scheme. (B) Alphavirus and coronavirus scheme. Details are provided in text.

somes recognize only the exposed 5'-end of RNA as a binding site for the 40S subunit. Thus, the RNA must be fragmented as monocistronic messages; there can be no additional internal initiation sites. It has been incumbent upon other ssRNA positive-strand viruses, all of which as noted have continuous, nonsegmented genomes, to devise other mechanisms.

It is necessary, of course, for progeny virus RNA to be synthesized, which in turn dictates the need for a template complementary to virus RNA. Two steps are to be found, the first synthesis of cRNA from the parent vRNA, and the second, replication of cRNA to progeny vRNA. Picornavirus vRNA is blocked at the 5'-end by a protein, vPg, which is irrelevant to translation but seems to serve as primer for RNA synthesis. Moreover, this family at least uses two replicases, one for cRNA minus strand synthesis and one for vRNA manufacture.

Alphaviruses and coronaviruses (Fig. 5B) generate a - strand RNA by replication of the parental virus RNA which is immediately used as a template for synthesizing new + strand molecules. If the entire - strand is not replicated, then abbreviated, or subgenomic + strands functional as mRNA can be formed. This now offers new initiation sites for ribosomal protein synthesis. The figure oversimplifies. For alphaviruses the input RNA is translated directly in part; the subgenomic RNA is translated to products which do not overlap those translated directly. Also, at least three proteins are derived from a single mRNA by cleavage. All the subgenomic RNAs for coronavirus, on the other hand, actually overlap considerably. The translation products differ because either they are read in a different frame or they are read only from nonoverlapping portions of the mRNAs.

For the ssRNA nonsegmented viruses, temporal controls thus exist for the most part over RNA synthesis only. As this is involved in all stages of replication (replication complexes, translation complexes, and progeny genome synthesis), it is obviously the critical element in the orderly regulation of events. Presumably, virus polypeptides participate at the level of the replicase complexes or RNA binding.

Negative-strand ssRNA Viruses. The remainder of the ssRNA viruses have negative-stranded genomes. The arenavirus genome is in two segments, bunyavirus in three, and orthomyxovirus in eight segments. Marburg-Ebola viruses appear to be a single negative-strand RNA virus, but the paramyxovirus and rhabdoviruses are definitely so. In the isolated state, negative-strand RNA genomes are nonfunctional, and permissive cells do not possess the necessary synthesizing factors which can replicate the genome to functional mRNA. Thus, isolated RNA is not infectious. Rather, the virion must carry along appropriate enzymes for RNA replication, and does so as part of the nucleocapsid complex. Upon infection, then, the transcriptase is already in close proximity to its substrate, and primary transcription is readily accomplished. The product of primary transcription is mRNA, and processing proceeds as with any messenger.

As expected, considerable variation exists. The rhabdovirus vesicular stomatitis virus (VSV) finds three proteins associated with its RNA, one a structural protein (N) and two subunits of the virion transcriptase (L and NS proteins). Upon infection transcription is initiated when (1) the M protein is dissociated, and (2) possibly cell factors associate with the transcriptase complex. RNA synthesis begins at the 3'-end; the first chain produced is a 47 nucleotide long "leader" RNA exactly complementary to the 3'-end of unknown function (not capped nor polyadenylated). After skipping three genomic nucleotides the N gene mRNA is synthesized, followed by mRNAs for NS, M, G (the membrane glycoprotein), and L proteins, in that order. Each mRNA thus depends on synthesis of the previous transcript in sequence; transcripts are polyadenylated by a virion enzyme which recognizes a sequence at the 3'-end of each mRNA. Thereafter cleavage occurs, and the 5'-end of each mRNA is capped to become a functional eukaryote messenger. Thus the "promoter" site for each gene seems to be approximately 30 nucleotides from the end of the preceding gene. The final 59 nucleotides at the 5'-end of the genome does not become mRNA.

Replication of new VSV genomic RNA requires that full-length + strand RNA be synthesized. This in turn demands that polyadenylation cease so that a continuous strand of cRNA is yielded. The mechanism for suppressing intergenic modification is unknown but may involve the N protein. Nevertheless, the same transcriptase recognizes the 3'-end of the + cRNA; the latter appears to circularize so that the replication complex may continuously recycle as it produces progeny - strand genome RNA. The interaction of new M protein with progeny RNA at the cell membrane (see Viruses and Cell Membranes) may once again inhibit transcription, as was the case during initial infection, and signal assembly of progeny.

The segmented genome of influenza finds each segment coding for a monocistronic mRNA (except segment 8, which codes for NS_1 and NS_2, which may be overlapping genes). Parental genome RNA is transcribed in the nucleus; the product is an incomplete transcript which is capped. Internal methylation is also carried out by host enzymes. As the genome is segmented, specific promoter sites and transcriptional controls are not a problem, and each transcript thus produced acts as mRNA. On the other hand, full-length transcripts of the parental virus RNA become templates for replicating progeny virus RNA segments. The full-length transcripts are primed by the transfer of a cap (7-methylated $GpppG^m$ or $GpppA^m$) plus approximately 25 nucleotides from cellular mRNA by a virion enzyme. Once primed, progeny genomes can be replicated.

Much of the RNA synthesis events are carried out by virion enzymes associated with the negative-stranded RNA viruses. As the amount of genome information is limited, this has required that many of the virus polypeptides are multifunctional. Thus, a given species of protein may serve both a structural role (as part of the ribonucleoprotein complex) and an enzymatic function. Further, many polypeptides undoubtedly serve more than one enzymatic function. Thus, when multiple enzymes are said to be associated with the virion, the conclusion cannot necessarily be derived that an equal number of polypeptides is involved. Nevertheless, the central role assigned to the virus in terms of RNA synthesis (with the cell responsible for protein synthesis) is reflected in the number of virus polypeptides serving these functions. Thus, influenza virus uses P1 and P3 in cRNA synthesis and P2 and NP for virus RNA synthesis; these four gene products alone are encoded in half the total genetic information.

dsRNA Viruses. Despite the fact that eukaryotic cells do not use dsRNA as a means of information storage or processing, this is a reasonably popular form of virus genome. In each instance the genome is segmented. The reoviruses and orbiviruses possess 10 segments, the rotaviruses 11, and a final unnamed group of related viruses (infectious pancreatic necrosis virus, oyster virus, infectious bursal disease virus of chickens, drosophilus X virus, and so on) have but 2.

The reoviruses are interiorized by phagocytosis, and replication is "activated" by lysosomal enzymes. These remove about half the protein of the outer shell, converting the infectious virus into the so-called subviral particle (SVP). The SVP is the replicating entity (Fig. 6); free dsRNA is never released into the cytoplasm, either because it would dangerously risk degradation by cell enzymes or because it would be too toxic to the cell.

The virion thus carries the various enzymes necessary to mRNA synthesis. These include an RNA-dependent RNA polymerase (transcriptase) capable of reading dsRNA, a nucleotide phosphohydrolase, pyrophosphate exchange, and a guanyl transferase and RNA methylase for capping the RNA product. The ssRNA is synthesized conservatively; that is, the dsRNA genome is transcribed to ssRNA which is capped and functions as mRNA leaving the genome intact (and in the SVP). The mRNAs are translated in conventional fashion to polypeptides. In view of the fact that genomic segments can be grouped by size (large, L; medium, M; and small, S) and numbered accordingly (L1-2, M1-3, S1-3), the polypeptide products can be genealogically identified by the Greek equivalents (polypeptides λ1-2, μ1-3, and σ1-3, respectively).

The genome must also produce ssRNA upon which complementary RNA strands are synthesized for progeny virions. The mRNA directed by parental SVPs serves this purpose and is catalyzed by a ssRNA $\rightarrow$ dsRNA transcriptase activity. The newly synthesized strand does not separate; it remains associated with the + strand as a progeny virus genome segment, and it occurs in association with newly formed progeny SVPs. The latter in turn remain active as the source of mRNA, with one important difference. Capping enzymes in progeny SVPs are masked, so that the mRNA produced by these structures are uncapped. This solves one problem for reovirus and creates another. It appears that the ss $\rightarrow$ ds replicase recognizes only capped mRNA; thus the potential overuse of mRNA for progeny genome formation at the expense of translation to virus polypeptides is avoided. However, uncapped mRNA is at a distinct disadvantage when competing with capped (e.g., cellular) mRNA. This dilemma in turn is resolved in two ways: (1) all mRNAs contain the sequence AUGG, complementary to the anticodon UACC in initiator RNA, and (2) more importantly, the translational machinery of infected cells become cap independent, presumably as the result of modification by virus polypeptides.

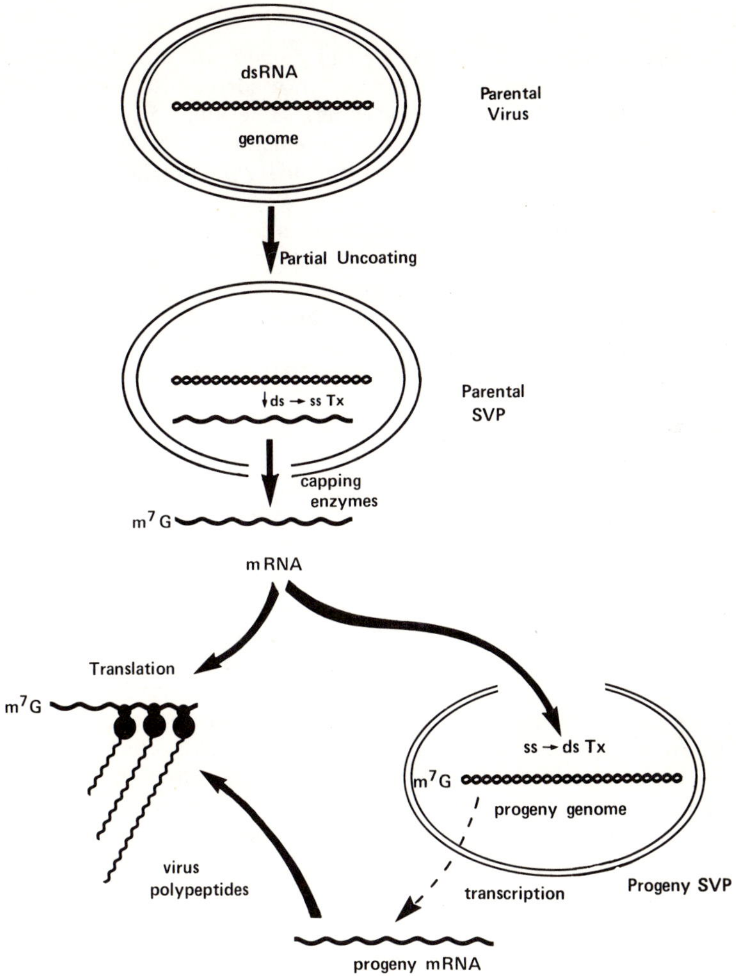

Figure 6 Replication scheme of dsRNA viruses.

For other dsRNA viruses replication schemes are not so clear. For example, infectious pancreatic necrosis virus produces four virus polypeptides encoded on but two genome segments by an unknown mechanism. Actually, intriguing problems with reovirus remain. For example, even though the virus assiduously avoids the relase of free dsRNA into the cytoplasm, these viruses are very potent interferon inducers. The mechanism of this event remains obscure.

Assembly

Assembly involves interaction of the primary products of intracellular replication—polypeptides and nucleic acid—in a series of events which result in the production of progeny virions. The steps involved are more or less complicated, depending on the complexity of the virus, and represent divergent considerations primarily as to whether the virus is naked or enveloped.

In general, the assembly of naked viruses is a sequential interaction of polypeptides to form subunits or precursors, followed at some point by association of progeny genomes. Frequently, polypeptides are further processed by posttranslational modifications after association into units. For example, adenoviruses, whose capsids consist of at least 12 polypeptides, form a shell as precursor to the final product. Structural proteins polymerize into intermediate subunits, which then are transformed into a precursor structure. DNA is inserted by virtue of its strong affinity via a specific binding site (approximately 150 base pairs from the left end). Picornaviruses similarly form a *provirion* containing viral RNA plus procapsid proteins. The latter are cleaved further after associating together, a reaction presumably triggered by the changes in configuration of the molecules as a result of this interaction. A favored mechanism used by many viruses is that one of the virus-specific polypeptides binds to progeny genomes at some point in the assembly process and prevents further transcriptional events. In this way transcriptase subunits are prevented from continued activity, thus interfering with assembly. In contrast to enveloped viruses, naked viruses tend to accumulate in the infected cell, sometimes forming crystals (e.g., adenovirus) and are released only by death and lysis of the spent host cell.

The assembly of enveloped viruses merits more detailed consideration in that a number of fundamental principles of virus-cell relationships and metabolic processes are involved. To prepare for egress from the cell the virus effects considerable modification of the cellular membrane. Assembly occurs at modified sites, and virogenesis follows. This process will be considered in detail.

Viruses and Cell Membranes

In addition to the various intracellular events occurring in the infected cell that lead to the production of progeny virus, a number of alterations in cell structure and function are found. Ultimately it is precisely these effects that determine the overall impact of the infectious process on the host. At the moment, however, we shall consider membrane alterations as a prerequisite to envelopment.

The surface of infected cells is considerably altered as the result of the infectious process. For the most part this is secondary to the appearance of newly synthesized polypeptides, most of which are encoded by the virus genome. This is true of cells infected with either naked or enveloped viruses.

The mechanisms involved bear on the basic processes by which glycoproteins themselves are synthesized. Most, if not all, surface components of virions are glycoproteins, and glycosylated proteins are synthesized in a manner distinct from other proteins. The latter are translated by free ribosomes, whereas the former are specifically translated by ribosomes bound to endoplasmic membranes, the so-called rough endoplasmic reticulum (RER). Synthesis of a polypeptide in this complex is initiated with a signal sequence of 15 to 30 amino acids which recognizes and establishes a membrane-ribosome junction, following which it is inserted through the membrane of the RER. The signal sequence thus appears as the amino-terminal sequence of the nascent polypeptide and acts as an anchor to establish the new membrane-associated virus polypeptide. The signal sequence is subsequently removed, and as continued ribosomal activity extends the length of the growing polypeptide, glycosylation occurs. The latter is accomplished via transfer of oligosaccharide moieties from a carrier molecule to asparagine residues. The immature glycoprotein then sequentially migrates to the smooth membranes, Golgi apparatus, and plasma membrane, during which time further processing of carbohydrate side chains takes place. In essence, then, virus glycoproteins are synthesized by the same route as proteins normally processed for export by the cell. The biological significance of immediate note is that virus-specific polypeptides appear throughout membrane systems of the infected cell, including the external surfaces. Here they serve as potential sites for virus maturation and, as they alter surface physical and antigenic properties, influence cell behavior.

The structure of the normal cell membrane ordinarily finds cell surface proteins anchored in or through a fluid lipid bilayer. The appearance of virus polypeptides seems to displace cellular proteins in whose place the virus proteins accumulate in "rafts" or "patches" (Fig. 7A). There has been considerable discussion as to whether a few normal cell antigens intermingle (e.g., H-2 antigens in VSV grown in mouse cells); if so, it occurs to only a very minor degree. Either the virus proteins possess an affinity for each other or are repulsed by cell proteins, accounting for their tendency to aggregate.

At the same time surface glycoproteins are appearing on the cell surface (which actually may occur very early in infection), other replication events are continuing, including the synthesis of M protein. This protein layers out on the cytoplasmic face of the cell membrane, probably because of its intermolecular affinity for the virus glycoproteins in the membrane (Fig. 7B). Two events follow. First, the accumulation of M protein is sufficient to induce polymerization as a

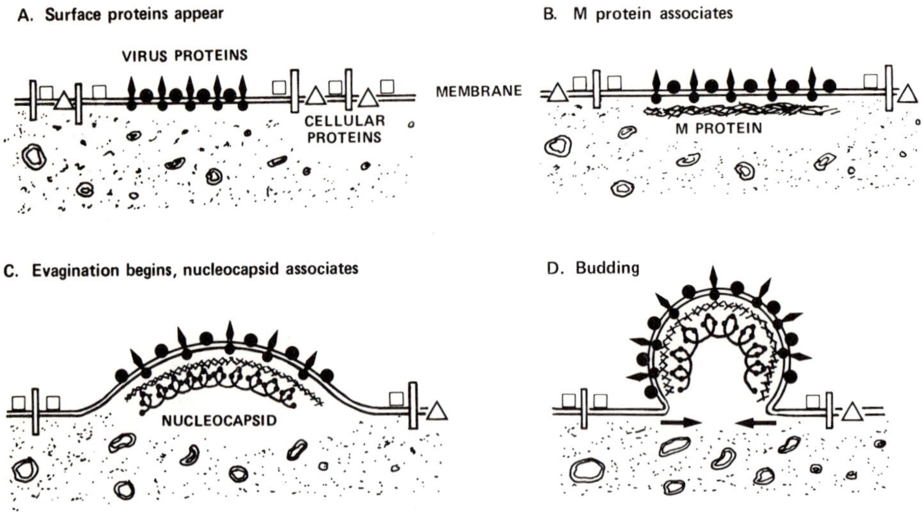

Figure 7 Diagrammatic representation of the budding process.

two-dimensional network, or matrix, along the interior of the membrane. This endows considerable rigidity to that portion of the cell membrane, a stimulus which seems to trigger evagination of the cell membrane. Perhaps this is mediated by microfilament elements of the cell cytoskeleton whose connections to the membrane are interrupted by polymerization of M protein, releasing that area of the membrane while still anchoring the surrounding cell surface. Second, the M protein in turn possesses intrinsic affinity for virus nucleocapsids so that these also become closely apposed along the layer of M protein. As the complex continues to evaginate M protein and nucleocapsids sequentially interact, ultimately the cell membrane pinches off while surrounding a layer of M protein wrapped around a nucleocapsid. A progeny virus has been formed by the *budding* process.

For some viruses the M protein is absent, but the primary interaction at the membrane is directly between the membrane glycoproteins and the nucleocapsid. This is true for both coronaviruses and togaviruses. During the formation of a bud approximately 400 host protein molecules are displaced and a virus envelope containing 800 molecules of virus polypeptides is substituted. As previously mentioned, the poxviruses do not bud but synthesize their own lipid-protein envelope. The immature enveloped virus is wrapped into a double-membraned sac by intracytoplasmic membrane, following which it migrates to the cell surface, where the outer membrane fuses with the cell surface and the progeny with its single membrane is released. Budding does not necessarily occur only at the outer membrane surface. Flaviviruses tend to bud into cytoplasmic vacuoles from whence they are released as the cytoplasmic vacuole fuses with the cell surface membrane.

It should be noted that, as the virus bud pinches off, cell fusion is again involved (1) to establish continuity of the virus envelope and (2) similarly to reestablish the integrity of the cell membrane. This again underscores the importance of membrane fusion events in the daily activities of the cell, and the dependence at least by enveloped viruses on such phenomena.

In addition to the means by which progeny-enveloped virions arise, the budding mechanism carries other implications for the infected cell. In that a budding virus represents a virion in situ, the same interactions found for the virus will occur with that region of the cell membrane. Thus, antibody directed against surface glycoproteins of the virion will also bind to the cell surface; if the appropriate conditions are met (e.g., complement activation, antibody-dependent cell-mediated cytotoxicity), the host cell may be lysed. If that particular virus will hemagglutinate erythrocytes, then these red cells are apt to bind to those receptors while still in the cell membrane, an event known as *hemadsorption*. Finally, if budding is to occur in close proximity to another cell of the same kind (which would usually be the case), then fusion of the two cells may be brought about (Fig. 8). The host cell obviously has receptors for the virus, as would the neighboring cell. Thus, attachment via receptors might occur before the budding process were completed. If this virus normally enters the new host cell by envelope-membrane fusion, then an intercellular bridge would be established. Cell fusion is actually a common event among cells in culture and is found with most enveloped viruses. This is recognized in infected cells as giant cells, multinucleate cells, or syncytia, since fusion can continue to occur until many cells have been incorporated. The significance is unknown; it may be an uneventful byproduct of virus replication. However, the process will continue, along with intracellular replication, in the presence of neutralizing antibody in the extracellular medium. Consequently it has been offered as an evasive action which viruses might pursue in the face of an humoral immune response by the host.

As alluded to earlier, not all new antigens which appear on the surface of the infected cell are virus specific. Rather, some are coded for by the cellular genome and appear because the infectious process either unmasks their presence or their synthesis is derepressed. One of the better known examples is the induction on the surface of infected cells of receptors for the Fc portion of immunoglobulin molecules. Normally, FcR is limited to B and (activated) T lymphocytes, mast cells, monocytes and macrophages, polymorphonuclear leukocytes, platelets, and tumor cells. Fibroblasts and epithelial cells infected

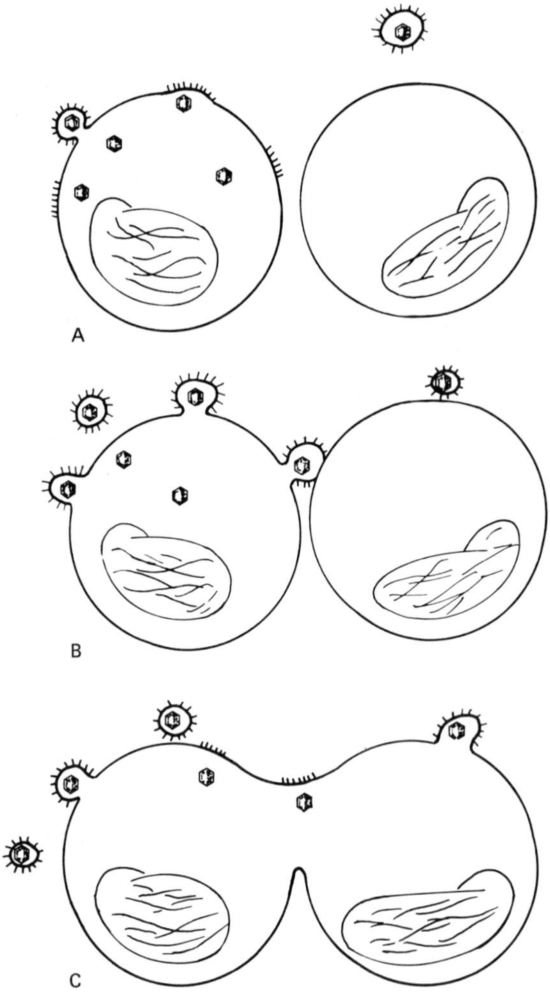

Figure 8 Mechanism of cell fusion during maturation of enveloped virus. (A) Virus budding from cell surface. (B) During budding, maturing virus has adsorbed to neighboring cell, establishing an intercellular bridge (C).

with HSV, CMV, and varicella zoster virus, however, also display FcR. That virus replication is responsible is demonstrated by the fact that phosphonoacetic acid (an inhibitor of HSV DNA polymerase) blocks the induction of FcR on infected cells but not on uninfected cells normally expressing them. In addition, of course, HSV infections also result in the appearance of a number of antigens, the HSV-specific glycoproteins gA, gB, gC, gD, and gE. The latter glycoprotein is, in fact, the FcR; its function, if it has one, is unclear. It may serve in a manner yet unidentified or, indeed, its ability to bind immunoglobulins may be crucial to the biology of HSV infections. For example, FcR will bind aggregated IgG, which will block complement-dependent and cell-mediated immune lysis, presumably by interfering with the approach of effector cells.

The envelopment process thus serves not only as a convenient means of exiting the infected cell for some viruses, but also represents an advantageous system to the host cell. For an already infected cell it costs little; the cell can afford to give up a considerable amount of its lipid membrane (as witness the phagocytic capacities of PMNs) and must synthesize only new replacement lipid. Viability is not interrupted. Finally, alteration of the external cell surface may act as an effective signal to the host that all is not well with the interior processes of the cell, and so appropriate responses may be initiated.

EFFECT ON THE HOST CELL

The most dramatic effect of virus infection on the cell is death. Given the finely tuned schemes of cellular metabolism designed to maintain viability, function, cooperative responses, and suppression of unneeded activities, it is easy to understand how significant interference with any cellular process might lead to catastrophe. However, as emphasized in the section on intracellular replication, a viable host must be maintained by a successful parasite, and the virus must carefully and with subtlety intrude into cell operations. It is the purpose of this section to explore how this is done.

The urgency with which host cell metabolism is suppressed by the virus reflects how rapidly the virus replicates to completion, plus the ultimate fate of the host cell. Thus, a virus whose replication cycle will consume a few hours and whose release will coincide with cell death will have to expropriate cell ribosomes, e.g., rather immediately compared to another virus which may not produce progeny for a matter of days or more, and the release of which may not jeopardize cellular well-being. In general, however, the presence of virus is reflected in a diminution of at least some of the normal metabolic processes in all cells and tends to correlate with the specific biosynthetic pathways required by the virus. Thus, it might be expected that DNA viruses would significantly influence host cell DNA metabolism, whereas RNA virus would only to a much lesser degree. That this is not strictly true owes in part to the fact that suppression of cellular protein metabolism, for example, will ultimately affect nucleic acid metabolism as well. After all, all but the substrates of nucleic acid biosynthesis are proteins—the enzymes and regulatory factors—so that impaired activity generally reflects takeover by virus factors or the diversion to virus requirements. Varicella zoster virus infections thus can be demonstrated to inactivate the host enzymes of pyrimidine nucleoside metabolism. An immediate transcription product of VSV acts in the nucleus to directly shut off host RNA metabolism. Exceptions exist, however. Adenovirus infections turn off host cell chromosomal DNA and histone synthesis but not mitochondrial DNA synthesis. Cytomegaloviruses actually stimulate cellular DNA synthesis, an event mutually exclusive of progeny virus production. In part this may mirror the desultory fashion in which this virus replicates and its tendency toward latent infection.

Probably all viruses interfere with host protein synthesis, but various agents accomplish it in different manners. St. Louis encephalitis virus infection exerts a barely discernible effect on total cellular protein synthesis, while poliovirus infection results in a rapid and abrupt cessation of translation. Infection with picornaviruses in general is accompanied by an increase in membrane permeability with a concomitant influx of sodium ions. In that virus replication is membrane associated it has been proposed that the insertion of virus components may open new Na^+ channels, may have a direct effect on the sodium pump mechanism, or may activate phospholipases which weaken the membrane. The increased intracellular tonicity shuts off protein synthesis directed by cell mRNA but not virus mRNA. As a result, virus polypeptides are preferentially translated. Hypertonic inhibition of cellular but not protein synthesis is a general phenomenon for many viruses, in fact, and has served as an useful investigative tool. The mechanism appears to be at the level of polypeptide chain initiation, in that the affinity of initiation factors is greater under hypertonic conditions for viral than for cellular messengers.

Additionally, other modifications occur in the poliovirus-infected cell to ensure preferential protein synthesis. Normally, eukaryote ribosome binding at the 5'-end of mRNA is enhanced by the presence of a methylated cap. However, certain virus mRNAs, such as poliovirus genomes, are not capped. As a result, upon infection poliovirus finds itself at a disadvantage in competing for ribosomes, since the latter bind with a higher efficiency to capped mRNA. In response, poliovirus replication products modify initiation factors during the course of infection so that the translational machinery now preferentially processes uncapped mRNA. The same process occurs during reovirus infections in which, it will be recalled, mRNA from parental SVPs is capped while mRNA from progeny SVPs is uncapped. Translation similarly becomes cap independent in this situation, so that parental mRNA is no longer bound by ribosomes but rather can be used full time as a template for synthesis of progeny genomes. These are two special situations, however, in that even other picornaviruses do not effect similar ribosomal alterations. EMC virus, for example, competes directly for initiation factors more successfully than cell mRNA by virtue of its intrinsically higher affinity.

It is not uncommon for one of the early proteins produced by viruses to be responsible for shutting off cellular protein synthesis. A number of viruses act even more immediately in that components of the virion itself are capable of inhibition. Vaccinia virus acts directly both at the level of initiation and, to a lesser extent, on elongation or release factors, to impair cell protein synthesis. Adenovirus inhibition is mediated in part by fiber antigen, the virion structure which in fact binds to the cell receptor.

A final example of virus modification of cell protein synthesis is offered by herpes simplex virus. First, a virion component acts directly to dissociate cellular polyribosomes. Subsequently, genome expression yields a product which degrades cellular mRNA. Ribosomes thus become available for virus mRNA and are recruited into virus polyribosomes without competition.

At the cellular level the consequences of metabolic interference are many. The impact which this will have on the host as a whole bears as much on the tissue site as it does on the rapidity or extent of cell death or injury. Thus, moderate damage to vital structures (e.g., CNS or cardiac tissue) may ultimately be more critical to the host than severe damage to a less important organ. Morphologically many viruses exert a significant effect on the infected cell. In large part this is mediated by alteration of phospholipid synthesis leading to dysfunction of cellular membranes. In addition, there is substantial disruption of cytoskeleton organization such that actin may be depolymerized (VSV) or virus products coat filaments (poliovirus and reovirus). In addition there is the occurrence of cell fusion as noted earlier. All these events lead to the cytopathic effects familiar to clinical virologists.

In addition to the direct damage suffered by the host cell as a consequence of virus replication, the pathogenesis of virus injury may be complicated by indirect influences on the function of other organ systems. The pathogenesis of influenza virus infection illustrates this point. Influenza virus infects and damages primarily the respiratory tract, and does so by direct injury to respiratory tract epithelial cells. As the result of impaired cellular metabolism, ciliary activity ceases. This abrogates the clearing function of the mucociliary apparatus which normally protects the lung from inhaled particulates. Accumulation of casual bacteria in the lower respiratory tract which normally would be eliminated is increased. Alteration of infected cell membranes by replicating influenza in some way allows certain bacterial species, especially staphylococci, hemophilus, and streptococci, to adhere, thus encouraging continued local multiplication. Additionally, progeny virus released from infected cells are also cleared less readily, encouraging dissemination toward terminal airways. In the alveolar sac the virus preferentially replicates in type 2 pneumonocytes (corner cells) of the alveolar epithelium and appears to destroy them. This deprives the alveolus of its source of surfactant, and the alveolus more readily collapses. Although cellular injury has no doubt evoked an inflammatory response, influenza virus is also capable of productively replicating in alveolar macrophages, so this serves to deliver fresh permissive cells to the infected site. Peripheral blood lymphocytes also possess receptors for the virus, but replication is abortive. The presence of virus, however, alters the cell's surface properties so that it loses its homing properties. Delayed hypersensitivity reactions are impaired, as are the number of circulating lymphocytes. Further effects on normal host immune functions are manifest by abolished monocyte chemotaxis, and particularly by the deleterious effect on polymorphonuclear leukocyte phagocytosis. Consequently the respiratory tract suffers not only directly the effects of virus replication but is also rendered vulnerable to superinfecting bacteria, a not inconsequential event in the natural history of influenza infections. Moreover, significant systemic symptoms are seen even though the virus infection is limited to the respiratory tract. Almost certainly these distant manifestations are secondary to locally released replication products. Suitable candidates for such factors include dsRNA molecules originating either from replication complexes or genomic RNA itself, segments of which demonstrate considerable base-pairing. These are directly toxic to protein synthesis in the cell and, if present in sufficient concentrations, may cause cell death.

The immunosuppressive effects of virus infections are not unique to influenza. The original observation in this regard was that delayed hypersensitivity to *Mycobacterium tuberculosis* was abolished during measles virus infections. This probably reflects a direct effect of intracellular measles virus on T-lymphocyte function. Interference with immune function, however, is found with many virus infections and is the result of diverse mechanisms which may involve all aspects of host immune responses. To illustrate: Sendai virus abrogates the ability of macrophages to kill bacteria, because phagosome fusion to lysosomes is inhibited. Avian leukosis virus infection of bursal-derived lymphocytes, at least in vitro, will selectively interrupt the normal

IgM to IgG switch in antigenically stimulated cells. Bovine diarrhea virus, on the other hand, suppresses function of both B cells (decreased synthesis of IgG as well as IgM) and T cells. HSV infections of lymphocytes cause these cells to lose both nonspecific (phytohemagglutinin) and specific (antigenic stimulation) blastogenic responses. Cytomegalovirus, conversely, increases susceptibility to superinfection and probably augments the chronicity of its own infectious process by inducing a repressor cell response, thereby effecting an overall decrease in immunoreactivity. A number of other examples could be cited, but it should be apparent that the immune functions influenced by virus infection may be manifold.

In the previous section surface alterations of the infected cell were discussed. A most important consequence of this phenomenon is that the cell or surface elements thereof become subject to immune injury as the result of immunologic reactions directed against these antigens (Fig. 9). Because membrane antigens differ from those normally present the infected cell is histoincompatible. Second, antigens released extracellularly and capable of stimulating an humoral immune response may also be fixed in the cellular envelope, so that antigen-antibody reactions may ensue. In many instances, then, it is actually immune reactions and not virus replication per se which damages or destroys the infected cell.

Cellular immune reactions are similar to those of transplantation rejection and involve direct attack on the virus-infected cell with membrane antigenic alterations. Such reactions are probably limited to the site of infection and are manifested as inflammation and tissue destruction of the involved area. It is entirely possible that these and other immunologic attacks on the host's own cells form the basis of "autoimmune" disorders involving similar but uninfected cells, but this possibility is only speculative. Immunoglobulins reacting with virus-specific antigens on cell surfaces carry additional implications. First, small amounts of bound Ig may induce "capping," in which the antigen-antibody complexes migrate to one pole of the cell and are interiorized or shed. This process denudes the surface of the infected cell of virus antigens, and the cell may once again appear innocent. Second, bound immunoglobulins may be capable of initiating complement-mediated or cell-mediated cytotoxic reactions. Such reactions may, of course, eventuate in the lysis of the participating cell. Further, an additional class of immune reactions precipitated by virus infections finds itself based on the fact that not only does the host cell shed progeny virions but it may also secrete considerable quantities of virus polypeptides as such. These molecules are antigens, of course, and secretion has been demonstrated to occur during the course of many virus infections. On the one hand, such antigens may be viewed as protective to the virus itself in that these soluble proteins may

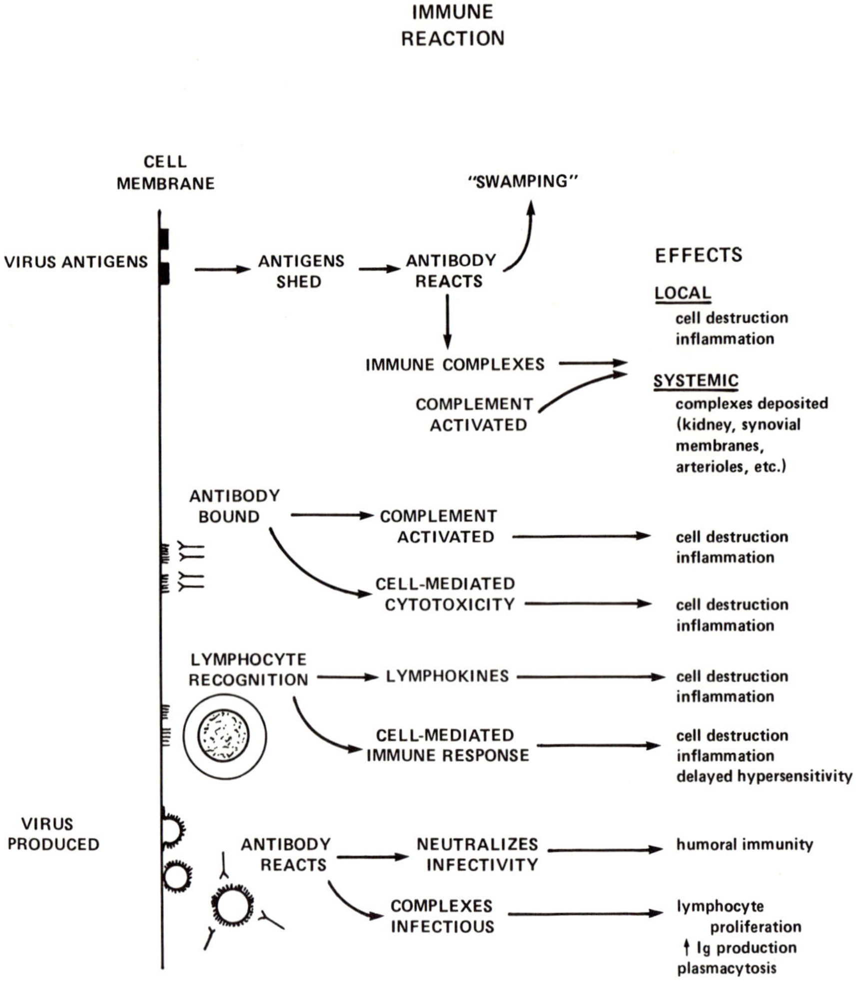

Figure 9 The infected cell surface and immunologic reactions. (See text for details.)

combine with environmental antibodies or immune cells and protect infectious virus by a "swamping" effect. Second, many of these virus polypeptides may be taken up by neighboring cells and, if intrinsically capable of such activity, suppress cell biosynthesis in these as yet uninfected cells. Finally, the release of antigens into the environment raises the possibility of immune complex formation which, depending on the relative amount of specific antibody present, may result in the focal precipitation of these complexes or the release into body fluids of soluble complexes. Either event probably has considerable significance for the pathogenesis of virus diseases mediated by locally precipitated immunoreactive complexes (e.g., on mucosal surfaces) or by circulating immune complexes.

A final sort of immunologically based virus injury is found in situations in which virus antibody complexes remain infectious. The mechanism is not known—it may be that virus and antibody can be dissociated intracellularly—but the phenomenon may be responsible for widespread tissue injury. The unchecked replication and release of virus antigens stimulates a progressively increasing antibody response which in some situations leads to unchecked lymphocytic proliferation, infiltrations, hypergammaglobulinemia, and often, autoimmune disorders. This is the mechanism of such diseases as lymphocytic choriomeningitis and lactic dehydrogenase virus infections in mice, Aleutian disease of mink, and equine infectious anemia of horses.

The pathologic features of virus infections thus depend on the overall effect of virus replication and host reactions on the physiologic function of the various tissue sites involved. The present discussion has centered exclusively on acute infectious processes, those in which virus replicates and is accompanied by cell damage. Acute disease is, in fact, the exceptional outcome of the infectious process. First, most newly acquired infections are inapparent unless they represent "accidental host" situations (e.g., rabies, herpes B virus, Lassa-Ebola-Marburg viruses in man). Second, if infection is defined as the presence of virus in the host, then it is probable that viruses remain in residence in the given host in a nonreplicating state for prolonged periods of time. This phenomenon is known as *latency* or *persistence* and will be dealt with in the next section. Finally, a very important consequence of virus infection is cell *transformation*, a putative equivalent of the malignant state. This subject is considered in other chapters.

PERSISTENCE

Extracellular virus is at risk of random degradative forces which cannot be anticipated in the adaptive sense or counteracted in any active way. Intracellular virus, on the other hand, is protected from such hazards and located at the optimal site for preserving and reproducing its genetic message. Active replication, however, sooner or later threatens the virus with the prospect of producing progeny virus, metabolic damage to the host, and inevitable release to the extracellular milieu once again. Assuming that one round of replication in the future, no matter when, is preferable to the likelihood of imminent destruction, it stands to reason that the preferable course is preservation of the viral genetic information in an intact but silent condition. Presumably this characterizes the selective pressures which have lead to the establishment of the persistent state as a part of the natural history of many virus-host cell relationships.

The basic problem in developing such a relationship is that the virus must have some mechanism by which replication (i.e., gene expression) is not suppressed in a permanent manner. While cellular nucleic acids may be stored in various compartments, especially the mitochondrium and the nucleus, free and autonomous nucleic acids have a finite life span. Thus, persistent virus genomes must be secreted away in protected sites. Second, the information must be accessible and recallable by some likely stimulus, usually either a physiologic change of state of the cell or reinfection by a related virus. If the entire genome is salvaged in the process this is known as *re-activation*; if only a portion of the genome is reactivated this is recognized as gene or *marker rescue*.

The mechanisms by which viruses are able to meet these conditions for persistence are many, and numerous clinical examples exist which testify to the success and the significance of the phenomenon. Many of the observations, however, are derived from various in vitro models of chronically infected systems in which viruses recovered from long-term cultures have been characterized. Consequently, in many instances the in vivo applicability of persistence mechanisms may be unknown. They nevertheless provide insight into the potential of viruses to persist and how this might relate to disease.

Defective-interfering particles are found with virtually all viruses studied. Some viruses may, however, have defective phenotypes (as opposed to the genotype) which contributes to persistence. The human disease subacute sclerosing panencephalitis (SSPE) is a late sequela of measles virus infection. The virus recovered from infected tissue is defective in that infectious virus is not normally produced, at least from infected tissue. Although the genome of these viruses is somewhat larger (by RNA content) than "normal" or *wild-type* virus (possibly suggesting recombination with another virus) the virus is apparently defective in its final product. Intracellular events suggest active replication, and the defect appears to be in maturation. Mutants produced in the laboratory have been characterized which have defective M proteins, so that one explanation might be that the budding process is deranged.

Measles virus also offers an example of a different form of persistence which, in this instance, depends on the host cell. It has been recognized for some time, and mentioned previously in this chapter, that viruses may infect leukocytes. Since these cells migrate throughout the organism, this offers the opportunity for viruses to disseminate intracellularly, circumventing extracellular defense mechanisms. Further, however, there is evidence that lymphocytes may serve as reservoirs for certain viruses. Uncommitted T lymphocytes are relatively inactive; upon stimulation by a specific antigen their metabolism is considerably enhanced. This increased activity also extends to RNA viruses. In the resting state lymphocytes are nonpermissive but the virus genome is preserved for long periods. Upon stimulation, then, virus replication is also turned on, and the cell begins to produce virus. As lymphocytes are very long-lived cells, this mechanism may preserve RNA virus genomes for significant periods of time. The relationship of DNA viruses may be somewhat different. CMV, for example, can persistently infect human lymphocytes but, rather than being quiescent, actively replicates. Replication is regulated such that an equilibrium is established between the rate of progeny virus being released and the growth of uninfected cells. In large part this is probably related to the controls over CMV replication exerted by the cell itself, in that replication requires a host cell function associated with the S phase of the cell cycle.

The physical state of virus genomes in many kinds of persistent infections is unknown, especially for RNA viruses. As previously mentioned, DNA viruses have recourse to chromosomal integration (often as multiple or tandem copies of the genome) but must be duly cautious that both integration and excision can be ensured. In addition, DNA virus genomes may be established as episomes and, in fact, may exist as an independent replicon. In this way multiple copies of the genome may be replicated. The advantage to the host cell is that quiescent genomes are obviously being held in check by repressors of some sort, and the latter will also act on superinfecting wild-type viruses. If their inclination might otherwise be to productively replicate and to harm the host cell, the same repressors should act on them and protect the host.

Integration of RNA retroviruses via a DNA intermediate is discussed separately. An intriguing extension of this idea has been presented for other RNA viruses. Here, it has been suggested that in cells coinfected with retroviruses the reverse transcriptase (RNA-dependent DNA polymerase) might also function to synthesize DNA copies of other RNA viruses. Such mechanisms were proposed specifically for measles and respiratory syncytial viruses. This hypothesis has not been confirmed although it certainly remains an interesting concept.

In vitro cultures of most viruses have been established which can be maintained indefinitely, even though the particular virus is initially highly cytocidal. In many instances these persistently infected cultures represent an equilibrium between permissive replication and cell multiplication. One cell factor which seems to be instrumental in maintaining the balanced state is interferon (see Chap. 3). In its simplest form, replication stimulates the release of interferon, which in turn discourages further replication. The cessation or diminished rate of replication decreases interferon production, favoring renewed replication. This reciprocal effect ultimately favors either a cyclic or steady-state condition and persistence of the infectious process. The clinical significance of this process probably more likely involves interferon as one of a number of cellular and immune factors which coordinately limit virus replication.

Virus attributes have also been implicated as contributing to persistent infections. In addition to the previously mentioned DI particles another universal product of prolonged and persistently infected cultures are *temperature-sensitive* (ts) mutants. These variants contain a biochemical lesion which inhibits productive replication at one temperature but which is normally functioning at (usually lower) other environmental temperatures. Thus, for example, mammalian ts virus replication is inhibited at 37°C but not at 33°C. The temperature-sensitive lesion may be in one or more cistrons of the virus, so it is possible that some genes might be expressed (and affect the host cell or interfere with the replication of superinfecting wild-type virus). It should be emphasized that ts mutants appear spontaneously and regularly in cultures infected with wild-type virus, and have been found for many different viruses (reovirus, measles, VSV, RSV, Sindbis alphavirus, and Sendai virus). In general, ts populations are heterogeneous with different lesions in various virus genomes. Many lesions in reoviruses have been characterized as extragenic suppressor mutations (gene X product suppresses the expression of gene Y), whereas a natural ts mutant of VSV produces an altered NS polypeptide.

The significance of ts mutants in the pathogenesis of disease in the intact host other than as a persistence mechanism is not yet fully appreciated. Whereas at first glance it might seem that a virus incapable of replicating at the temperature of the intact host's cell would be benign, this is not necessarily the case. A ts strain of VSV, for example, causes a slowly progressive spongiform degeneration of the CNS in BALB/c mice (but is avirulent in Swiss mice). In part this may owe to the fact that the genome is still partially expressed; products thereof thus may be active in affecting cellular metabolism or eliciting immune responses.

A number of other mechanisms exist which encourage persistent infection in specific examples. Adenoviruses may infect nonpermissive cells and persist for prolonged periods; that the genome remains fully functionable can be demonstrated by subsequent rescue with SV40 helper virus. Visna virus in sheep undergoes periodic antigenic alterations, circumventing specific immune defenses and producing progressive CNS disease with cyclic exacerbations. Certain enveloped viruses similarly masquerade by forming *pseudotypes*, wherein nucleocapsids of one virus buds through and acquires the envelope of a different virus.

In the final analysis the persistent state represents in most instances a finely balanced combination of virus and host factors which leads to latency or continued levels of tolerable activity. Recrudescence or cure may eventuate, depending on whether the host factors predominate. Clinically these are all reflected as susceptibility factors, which are not always easily discerned from virus virulence factors, with which there exists a reciprocal relationship. One mediator system of host stress, such as trauma, fever, or ultraviolet light, for example, is the prostaglandins. These have been shown to enhance HSV replication, presumably by increasing intracellular levels of cyclic adenosine 3',5'-monophosphate (cAMP) which directly stimulates DNA virus replication. Prostaglandins are but one factor controlling replication, however, and are also known to enhance replication of CMV, adenoviruses, and retroviruses. Herpesvirus latency is in some way encouraged by the extracellular presence of specific IgG antibody. It has been proposed that the glycoprotein gE, or FcR discussed previously, serves to bind immunoglobulins and thereby signal their presence. More recent evidence, however, suggests that antibody is not essential to maintaining latency, although this does not rule out its role in initiating the condition.

Silent reservoirs thus participate in a major way in the natural history of viruses. As genome characterization becomes more precise and available the stability of virus nucleic acids will become increasingly apparent. For example, the H1N1 influenza recrudescent in 1977 has been found to be virtually identical to that strain first identified in 1950. If it is assumed that molecular evolution is an inevitable concomitant of replication, then the existence of identical genomes separated by a quarter of a century implies molecular hibernation. Revelation of the cellular site and host await the proper ingenious approach. In the meantime the concept of virus persistence may well be invoked as an explanation of many chronic disease processes.

CONCLUDING REMARKS

This chapter has attempted to present the virus infectious process as a dynamic but controlled state of biochemical activity of foreign nucleic acids as they parasitize the host cell. Rather than being a frantic incursion in which the virus attempts to urgently exploit the cell's biosynthetic capabilities, in many instances a long-term relationship with the host is to be anticipated. To be sure, the immediate early goal is replication. During this process various host barriers (both active and passive) to the replication cycle must be circumvented. Ultimately, however, the aggressive character of the virus and the defensive nature of the host reach the equilibrium necessary to avoid the extinction of one by the other. The biology of this equilibrium implies numerous attributes and features not discussed, such as virulence or attenuation of the virus and resistance or susceptibility of the host. It is the need for this equilibrium which results in the finely controlled biochemical maneuvers pursued by the protagonists and which have formed the subject matter of this chapter.

The basis of life is genetic nucleic acid. The virus-host relationship is a fundamental interaction of two "species" of nucleic acid. If, however, we view phenotypes as the means by which both genotypes armor themselves against the environment, then four permutations of the host-parasite relationship can be considered:

1. Virus phenotype-host phenotype: the epidemiology of infection in a population; host defense mechanisms (e.g., antigenicity)
2. Virus genotype-host phenotype: replication cycles and persistence
3. Virus genotype-host genotype: integration mechanisms, control of gene expression
4. Virus phenotype-host genotype: inheritable host susceptibility factors (e.g., immune response genes, HLA types, racial and gender predilections)

The genotype can adapt, guided by its intrinsic propensity to mutate and by selective pressures exerted on the phenotype. In a sense, then, the mammalian host deals with viruses in much the same way it must its "own" genes, which overduplicate, transpose, mutate, repress, express, repair, transform, and help and harm the organism. A virus may transcend the individual host's ability to control it. In a population, nevertheless, an equilibrium will be established just as is the case with harmful genes.

SUGGESTED READINGS

Carter, W. A., and E. DeClercq. Viral infection and host defense. *Science 186*:1172-1178, 1974.

Choppin, P. W., and A. Scheid. The role of viral glycoproteins in adsorption, penetration, and pathogenicity of viruses. *Rev. Infect. Dis. 2*:40-61, 1980.

Dubois-Dalcq, M., and B. Rentier. Structural studies of the surface of virus-infected cells. *Prog. Med. Virol. 26*:158-213, 1980.
Flint, J. The topography and transcription of the adenovirus genome. *Cell 10*:153-166, 1977.
Huang, A. S., and D. Baltimore. Defective viral particles and viral disease processes. *Nature (Lond.) 226*:325-327, 1970.
Kohn, A. Early interactions of viruses with cellular membranes. *Adv. Virus Res. 24*:223-276, 1979.
Sissons, J. G. P., and M. B. A. Oldstone. Antibody-mediated destruction of virus-infected cells. *Adv. Immunol. 29*:209-260, 1980.
Sissons, J. G. P., and M. B. A. Oldstone. Killing of virus-infected cells: The role of antiviral antibody and complement in limiting virus infection. *J. Infect. Dis. 142*:442-448, 1980.
Weinberg, R. A. Integrated genomes of animal viruses. *Ann. Rev. Biochem. 49*:197-226, 1980.

chapter 2

Immunopathology in Viral Disease

Immune Enhancement of Dengue Virus Infection

SCOTT B. HALSTEAD

John A. Burns School of Medicine
University of Hawaii at Manoa
Honolulu, Hawaii

Concepts of viral immunopathology have grown in direct proportion to the dimensions of modern immunology. Viruses interact with the immune system in three major ways: as antigens which complex with determinant-recognition sites on immunoglobulin molecules or immunocompetent cells, as parasites of cells which possess immune function, and as molecular precursors to autoimmunity. When any of these processes result in a functional change perceived as abnormal, it has become common practice to describe the phenomenon as "viral immunopathology."

The first category of viral immunopathology includes events so familiar it is difficult to think of them as immunopathologic. This is because the interactions of virions, virion subunits, and nonstructural viral antigens with antibodies or cell recognition sites normally stimulate the production of many of the inflammatory signs and symptoms we recognize as viral "disease." Many persons intuitively believe that viral illness is largely due to the destruction of cells by the process of virus replication. Closer inspection shows that symptoms often coincide with the initiation of the process of virus elimination by the immune system. In the opinion of some workers, an infection is immunopathologic only when the immune response is an obvious part of the disease process. By this criterion, immune complex disease due to lymphocytic choriomeningitis (LCM) virus in mice, Aleutian disease of mink, and hepatitis B vasculitis in humans are by consensus immunopathologic conditions. Cell-mediated immune responses also may produce immunopathologic disease. By adoptively transferring primed T lymphocytes into immunosuppressed hosts, it has been possible to demonstrate that lymphocytic choriomeningitis virus, normally not cytocidal for mouse cells, is quickly eliminated in the presence of immunocompetent cells. Active cellular immunity converts LCM from a benign to a fatal infection. Might not the same phenomenon be at work in measles, rubella, mumps, and chickenpox, systemic infections in which symptoms appear late and coincide with the initiation of the humoral immune response?

The second category of viral immunopathology has been demonstrated principally in experimental animals. The depression of cell-mediated responses following measles infection in humans is a well-recognized, if poorly understood, example of altered immune function caused by virus infection of immunocompetent cells. In experimental animal systems, virus infections may depress delayed hypersensitivity reactions, prolong allograft rejection, decrease in vitro lymphocyte reactivity to specific antigens, or decrease lymphocyte responsiveness to mitogen stimulation or mixed lymphocyte reactions. Viral infection may also affect the humoral immune response, producing elevations or depressions in levels of serum immunoglobulins, enhancement or suppression of specific antibody production, or suppression of the induction of high zone tolerance.

The role which viruses may play in triggering autoimmune phenomena has not been well delineated. Theoretically, autoimmune disease might result from the deletion of clones of suppressor cells or by a hapten effect in which the interface between viral and host cell membrane molecules form antigenic determinants which eventually result in the elaboration of clones of cells which react with only the host cell component.

A fourth immunopathologic mechanism, heretofore unrecognized, may be surprisingly common. This has been termed immune or antibody enhancement of viral infection. The phenomenon has two fundamental prerequisites: (1) the virus must possess the ability to replicate in cells of mononuclear phagocyte lineage (e.g., monocytes, macrophages, histiocytes, Kupffer cells), and (2) at the outset of infection the host must possess reactive but nonneutralizing antibody. Nonneutralizing antibody may form infectious complexes which are differentially taken up by mononuclear phagocytes, resulting in an increase in the number of infected cells and enhancement of the resultant disease.

In the mid-1960s, it was established that a severe syndrome caused by dengue viruses, dengue hemorrhagic fever (DHF), was significantly associated with a second infection in an individual previously immune to one of four dengue virus types. At first, DHF appeared to be a naturally occurring instance of some sort of "viral hypersensitivity." Subsequent studies have revealed a simple and elegant explanation—antibody enhancement of viral infection. Human dengue infection may serve as a useful experimental model for realizing the long sought goal of unveiling virus disease pathogenesis at the macromolecular level. In the molecular domain, a large number of vertebrate viruses will undoubtedly share common pathogenetic mechanisms. The remainder of this chapter explores the immunopathogenetic mechanisms in dengue hemorrhagic fever and the implications of the immune enhancement hypothesis for other viral diseases of humans.

THE AGENT

Dengue viruses belong to the genus *Flavivirus* in the family Togaviridae, a group of enveloped, single-stranded RNA viruses. There are presently 60 accepted members of the genus, all sharing one or more envelope group determinants which can be detected in the hemagglutination inhibition (HI) test. Among the flaviviruses there are four distinct viruses which are transmitted by *Aedes aegypti* and which produce the dengue fever syndrome in humans. These viruses can be distinguished by the plaque reduction neutralization test using antisera raised in naturally infected human beings or experimentally infected susceptible monkeys. Antisera raised in mice using adjuvant and limited inocula can be used to type virus using the complement fixation (CF) test. Antisera which are type specific in the neutralization test are generally cross reactive when the same antigens are studied by fluorescence microscopy.

IMMUNE RESPONSES TO DENGUE INFECTION

Figure 1 describes the approximate temporal relationships between dengue infection and antibody responses. Infection of a *Flavivirus* virgin by dengue virus is referred to as a *primary infection*. This is followed by a classic primary-type antibody response; HI and type-specific neutralizing (Nt) antibodies are usually detectable at the end of the viremic period and are of the IgM class. Three or more weeks after onset of symptoms, CF antibodies appear. This signals the appearance of antibodies of the IgG class. In tests in vitro, anti-dengue IgM does not fix complement, while IgG does so readily.

Primary infection results in lifelong circulation of HI, CF, and Nt antibodies and presumed lifelong homotypic immunity.

Dengue inoculation of an individual immune from a prior infection with a different member of the *Flavivirus* genus results in a *secondary infection*. This is accompanied by a secondary-type immune response characterized by an anamnestic antibody response to the determinants shared between the initial and subsequent infecting viruses. When

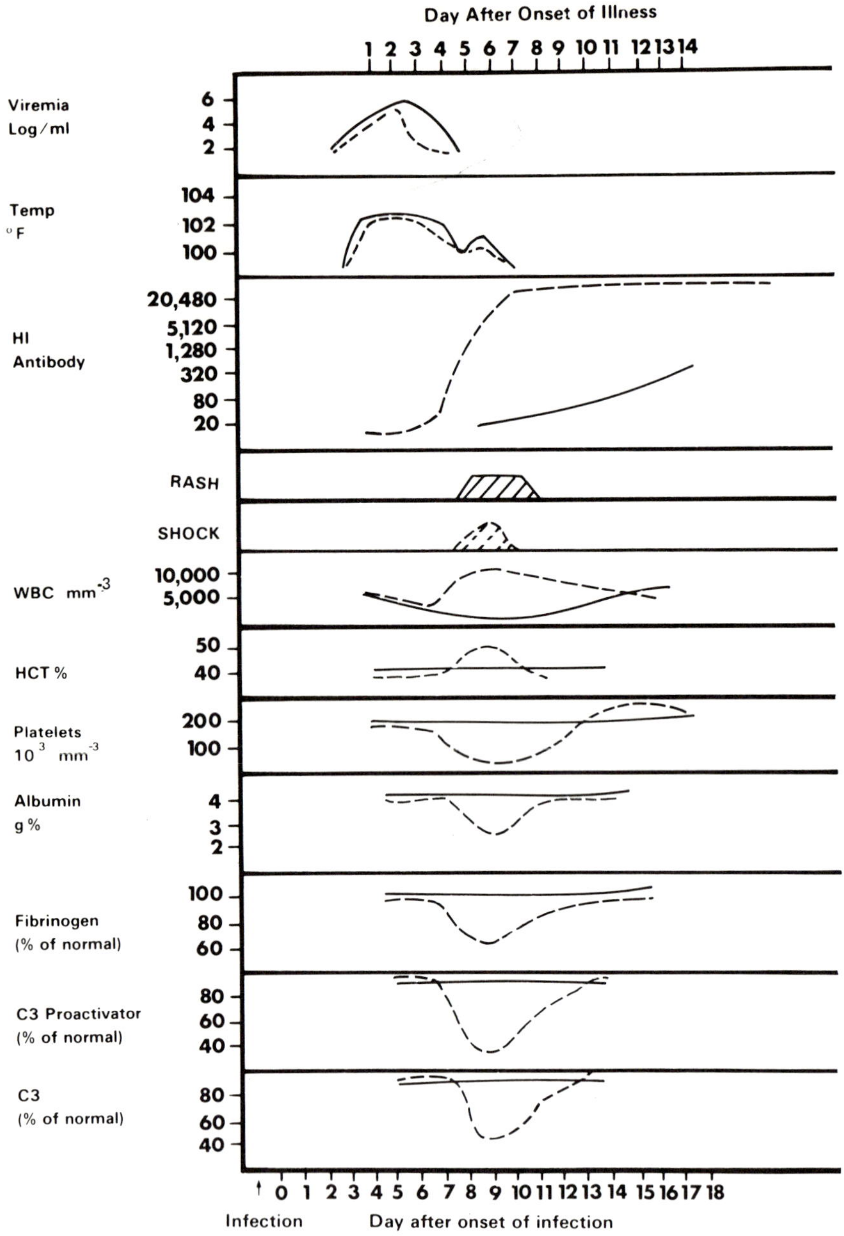

Figure 1 Comparison of the temporal appearance of selected clinical and laboratory abnormalities in classical dengue fever (—) and dengue shock syndrome (---). [After S. B. Halstead, Immunological parameters of togavirus disease syndromes. In *Togaviruses* (R. W. Schlesinger, ed.). Academic, New York, 1980.]

the two infections are members of the dengue subgroup, usually little or no antibody of the IgM class is detected during secondary infections. Secondary and primary infections can be distinguished by (1) the rate of production of antibody, (2) the amount of antibody produced, (3) the specificity of antibody, (4) presence or absence of CF antibodies, and (5) the presence or absence of anti-dengue IgG or IgM. Relatively simple criteria are sufficient to distinguish primary and secondary infections in most cases.

DENGUE SYNDROMES

Careful observation on dengue infections in children in Thailand has revealed a continuum of host response, from mild or inapparent infections to an illness characterized by shock and various bleeding manifestations. At the extremes, these are recognized as two syndromes—undifferentiated fever, or childhood dengue fever, and dengue hemorrhagic fever-dengue shock syndrome (DHF-DSS). The continuum of host response provides an important insight into the possible pathogenetic mechanisms of the severe end of the spectrum. It is logical to expect that DHF-DSS is an exaggeration of processes operating at subthreshold levels in mild dengue, rather than an entirely novel pathogenetic mechanism. The spectrum of host response observed in poliomyelitis is a better analogy for dengue disease than is the relationship of streptococcal pharyngitis to rheumatic fever. With this in mind, an examination of dengue syndromes is in order. The principal features of dengue fever and DHF-DSS are contrasted in Figure 1.

Dengue Fever

Clinical findings in mild, self-limited dengue infections vary with the age of the individual. In adolescents and adults, dengue infections are frequently accompanied by the dengue fever syndrome. While there may be distinctive clinical features to infections associated with each dengue virus type, there is as yet insufficient evidence on this point.

The features of the dengue fever syndrome are better described than those of pediatric dengue and will be given here for illustrative purposes.

After an incubation period of 2 to 7 days, fever suddenly rises to 103 to 106°F. This is frequently preceded by back pain and accompanied by a frontal or retroorbital headache and a transient macular rash. In the meticulous human volunteer studies conducted by the U.S. Army in the Philippines in 1923 and 1929, it was established that viremia begins slightly before onset of fever, and as viremia ends, so does fever. Myalgia or bone pain occurs shortly after onset. During days 2 to 6 after onset of symptoms, nausea, vomiting, anorexia, taste aberrations, generalized lymphadenopathy, and cutaneous hyperesthesia may develop.

Coincident with defervescence, a generalized morbilliform maculopapular rash appears, which spares the palms and soles. About the time of the second rash, the body temperature may briefly rise from previous values, resulting in the biphasic temperature curve.

Hemorrhagic phenomena have often been associated with the dengue fever syndrome. Usually they are minor. Epistaxis, gum bleeding, petechiae, and purpuric lesions which vary in frequency from outbreak to outbreak, may occur at any stage of disease. Gastrointestinal bleeding, menorrhagia, and bleeding from other organs are observed rarely, mainly in adults.

The hemopoietic system, heart, and brain are affected by dengue infection. Swift destruction of mature polymorphonuclear leukocytes is a regular feature of the febrile phase of dengue fever. This phenomenon is not well studied, but might be related to an early, generalized bone marrow depression described in some dengue-infected humans. In addition, T-wave changes and post-illness bradycardia are common. Psychomotor depression may be a cause of severe post-illness disability.

Children with mild dengue infections have a febrile illness accompanied by pharynitis and upper respiratory signs. Rashes, although less common, are as described for adults.

Dengue Hemorrhagic Fever-Dengue Shock Syndrome

This manifestation of dengue infection has a characteristic clinical progression. The period between mosquito bite and onset of fever is unknown, but is presumed to have the same range as in dengue fever. In children, there is a relatively mild first phase of illness with abrupt onset of fever, malaise, vomiting, headache, and anorexia with symptoms of pharyngeal and upper respiratory tract inflammation. Coincident with defervescence, there is a rapid deterioration in the condition of the child. The median day after onset of fever of the abrupt appearance of DHF-DSS is 4 days. On physical examination, the patient usually manifests cold, clammy extremities, a warm trunk, flushed face, and diaphoresis. The child is restless and irritable and complains of midepigastric pain. Often, there are scattered petechiae on the forehead and extremities, spontaneous ecchymoses, easy bruisability, and bleeding at venipuncture sites. Respirations are rapid; circumoral and peripheral cyanosis may be observed. The pulse is weak, rapid, and thready. If the physiological condition stabilizes at this point and the child does not develop hypotension or narrow pulse pressure, the case is considered dengue hemorrhagic fever *without shock*.

If the pulse pressure becomes abnormally narrow (20 mmHg or less) or the systolic or diastolic pressures are low or unobtainable, the case is defined as DHF with shock or *dengue shock syndrome*. It should be emphasized that children who are not clinically in shock may be hypovolemic and are often very ill.

Coincident with the period of acute hypovolemia or soon afterward, the liver may become palpable two or three finger breadths below the costal margin. (The liver is normally firm and nontender.) Gross ecchymoses and/or gastrointestinal hemorrhage may be seen in as many as 10% of DHF patients; this is usually associated with shock syndrome and is seen after the onset of shock.

Children who survive a 24 to 36 h period of crisis recover fairly rapidly. T-wave changes during illness and bradycardia after illness are often seen.

Laboratory Findings

The principal laboratory features of DHF-DSS are those associated with (1) *increased vascular permeability*. The diagnosis cannot be made unless the hematocrit or hemoglobin is elevated. Chest x-rays show pleural effusion, more notable on the right side. The defect in the vascular system has a characteristic molecular size, since relatively more albumin leaks into interstitial spaces, while larger macromolecules, such as immunoglobulins, are retained in circulation. (2) *Mild liver damage*—SGOT and SGPT values are elevated. (3) *Fever and infection*—children have mild to severe metabolic acidosis. (4) *Abnormal hemostasis*—thrombocytopenia is always present, as is a prolonged bleeding time. About one-third of shock cases have a prolonged prothrombin time. Half these patients exhibit prolonged partial thromboplastin time, and reductions in factors II, V, VII, IX, and XII. Fibrinogen levels are normal or low-normal, and in severe cases there may be elevation in fibrinogen degradation products. Bone marrow aspiration shows maturation arrest of megakaryocytes during the early acute state of the disease. (5) *Consumption of complement proteins*—shortly before and during shock, blood levels of Clq, C3, C4, C5-8, and C3 proactivator are depressed and C3 catabolic rates elevated. Further, there are reduced blood levels of serum carboxypeptidase B, a C3 inactivator. Circulating immune complexes have been detected during the febrile phase of DHF-DSS.

All the laboratory abnormalities described vary directly with the severity of illness. The activation of these inflammatory systems implies a parallel increase in substrate. In secondary infections, the degree of activation of Clq parallels disease severity, implying that the quantity of antibody *and* antigen produced also increase with disease severity.

Pathology

Although dengue viruses have been a significant cause of death in many Southeast Asian countries for over two decades, there are relatively few published studies on dengue pathology. Many reports do not include virologically studied cases. Ultrastructural studies are limited to a few tissues only. From the standpoint of pathogenesis, there is meager morphologic evidence revealing the site of dengue replication in humans. The central findings can be summarized as follows.

There is gross and microscopic evidence of generalized leakage of proteinaceous fluid and red blood cells from the vascular compartment. This is not accompanied by visibly damaged endothelial cells, or inflammation near basement membranes. Intravascular thrombosis is not observed.

Interesting changes are found in the lymphatic and mononuclear phagocyte (reticuloendothelial) systems.

1. Necrosis and hyalinization of Kupffer cells.
2. Focal necrosis of liver cells. In 81% of a series of 100 autopsies in Thailand, Councilman-like bodies were found and there was swelling and hyaline necrosis of Kupffer cells. These were conspicuous in the vicinity of foci of necrotic hepatocytes. Necrosis of Kupffer cells was sometimes seen without necrosis of hepatocytes.
3. Marked bone marrow depression occurs during the febrile state of illness. In most instances there is evidence of recovery at the time of death. Bone marrow macrophages demonstrate extraordinary phagocytic activity involving all elements of the blood: erythrocytes, lymphocytes, and platelets.
4. Profound depression and destruction of lymphocytes in thymus and in the T-dependent areas of the spleen and lymph nodes.
5. Reticulum hyperplasia in germinal centers in the spleen, marked phagocytosis of necrotic splenic lymphocytes and erythrocytes, increased plasma cells in Peyer's patches with necrotic and hyperplastic germinal centers. Some of these findings may be related to secondary-type antibody responses.
6. An interstitial pneumonitis which is largely due to the trapping of formed blood elements, including megakaryocytes, in alveolar capillaries.
7. Finally, there is evidence from microscopic, ultrastructural, and fluorescent antibody studies of mild proliferative glomerulonephritis with immune complex deposition. In one study of surviving patients biopsied at intervals to day 24 after onset of fever, a progressive, mild glomerulonephritis was observed. Despite evidence of involvement of glomeruli, urinary abnormalities are usually rather trivial.

Pathogenesis

A limited number of attempts to identify the site of dengue infection in humans have been made. No specific fluorescence was observed in frozen sections or imprints of fresh thymus, lung, lymph nodes, kidney, adrenal, skin, heart, or liver in 21 Thai autopsies. In a single case, fluorescence was noted in "large lymphoid or reticulum" cells in the sinusoids.

Renal biopsies from 20 cases of DHF-DSS in Thailand failed to reveal antigen by fluorescence microscopy. However, 40 to 50 nm dengue virus-like particles were visualized in glomerular monocytes and mesangeal cells in each of the 12 cases studied by electron microscopy. In some instances, particles were in clusters or in crystalline arrays. Particles were not enveloped. They resembled structures described in dengue-infected mouse brain, mosquito cells, and Raji human B lymphoblastoid cells.

The best study to date on the site of replication in human tissues is a series of skin biopsies obtained from 53 Thai children, each with a confirmed dengue infection. Skin sections were studied by fluorescence microscopy. Specificity of staining was verified by blocking, adsorption of sera with dengue antigens, and use of nondengue antibodies. Areas of macular, papular, and petechial lesions and normal skin were biopsied. In 14 cases, cytoplasmic fluorescence was localized to mononuclear leukocytes found beneath dermal papillae, closely adjacent to but outside dermal capillaries.

Recently, again from Thailand, dengue virus has been repeatedly isolated from glass-adherent blood leukocytes obtained on the first day of hospitalization (day of defervescence) in a large series of children with DHF-DSS.

At death, patients are usually virologically sterile when organs are tested using suckling mice or tissue culture systems. To date, one strain of dengue 4 was isolated from the liver in a fatal case from Thailand, and dengue 2 viruses have been isolated from bone marrow, lymph node, lung, and liver from one patient each (two in Thailand; two in Singapore). Limited attempts to isolate viruses from tissues by inoculating diluted suspensions (to disassociate virus-antibody complexes), to inoculate immediate postmortem liver biopsies processed without freezing, and to detect dengue antigen by using tissue suspensions as an immunizing reagent have been negative.

As a possible explanation of the absence of either virus or viral antigens at death, various tissue suspensions from 9 of 11 fatal Thai cases were shown to contain neutralizing substances specific for dengue viruses.

Two other observations may be relevant to pathogenetic mechanisms. (1) Rates of recovery of virus from the blood of patients with secondary-type dengue antibody responses are lower on each day of illness in patients with fatal outcome compared with survivors. (2) Antibody titers in fatal DSS cases with secondary-type antibody responses are lower on each day of illness than in patients with or without shock who survive illness.

The possible significance of these phenomena receives comment below (see Discussion).

EVIDENCE FOR AN IMMUNOPATHOLOGIC ETIOLOGY FOR DHF-DSS

Epidemiologic Observations

In 1967, an analysis was published of serologic studies on DHF-DSS causes in Bangkok. This resulted in the so-called two-infection hypothesis of DHF-DSS. In fact, this was not an hypothesis, but an observation: shock in dengue infections was significantly correlated with a secondary-type antibody response. As an explanation, it was proposed that DHF-DSS was a "viral hypersensitivity" syndrome, mechanism unknown. The early evidence for an association between DHF-DSS and a secondary-type antibody response is summarized in Table 1.

The association of DHF-DSS with a second dengue infection was promptly confirmed in epidemiologic studies done in Bangkok in 1962 and Koh Samui Island in the Gulf of Siam in 1966 and 1967 (Table 2).

The observed correlations between secondary dengue infection and DHF-DSS resulted in what might be termed the acute immune complex hypothesis of the etiology of the syndrome. Simultaneous production of dengue antigenic material and dengue antibodies of the IgG class results in immune complex activation of complement by the classic pathway, resulting in the production of C3a anaphylatoxin. This, in turn, mediates vascular permeability through the release of histamine. Interactions with the blood clotting system were hypothesized.

It is now believed that the acute immune complex hypothesis does not fully explain major clinical and epidemiologic features of DHF-DSS. Any useful hypothesis of the pathogenesis of DHF-DSS must take into account a number of idiosyncratic phenomena associated with the syndrome. These are as follows.

Age

DHF-DSS in infants comprises a substantial proportion of total cases contributing to the characteristic bimodal age-specific hospitalization rate curve. The mode of infants less than 1 year of age are composed almost entirely of primary dengue infections, while children hospitalized with DHF-DSS in the group 1 year old and older are experiencing secondary dengue infections (Fig. 2).

Table 1 Association between Secondary-Type Antibody Responses and Severity of Dengue Disease among Bangkok Children's Hospital Study (Ages < 1-14)

	No. of cases of primary infection	No. of cases of secondary infection	Secondary infections as % total infections
All children[a]	165,794	125,728[b]	43.1
Outpatients, FUO[c]	33	61	64.9
Inpatients, FUO[d]	13	23	63.9
Inpatients, DHF, nonshock[e]	65	262	80.1
Inpatients, DSS[e]	6[f]	190	96.9

[a]S. B. Halstead, Immunological parameters of Togavirus syndromes. In *Togaviruses* (R. W. Schlesinger, ed.). Academic, New York, pp. 107-173.

[b]Data derived from nonhospital-based epidemiologic study.

[c]S. B. Halstead, et al. *Am. J. Trop. Med. Hyg. 18*:972-983, 1969.

[d]S. Nimmannitya, et al. *Am. J. Trop. Med. Hyg. 18*:954-971, 1969.

[e]S. B. Halstead, et al. *Yale J. Biol. Med. 42*:311-328, 1970.

[f]Includes 4 infants < 1 year of age.

The antibody status of mothers of infants with primary infection DHF-DSS has been studied in only a few cases. It is known for the period of study shown that virtually all women of childbearing age in Bangkok were immune to one or more dengue viruses. Mothers of infants with primary DHF-DSS who have been studied in each instance have been immune to multiple dengue viruses.

Sex

Most of the published DHF-DSS autopsy series have demonstrated a higher incidence of deaths in females than in males. The distribution of shock syndrome by sex has been studied carefully in one large hospital study. Sex differences were observed, but only in older children. Among primary dengue infections, regardless of age, slightly more males than females were admitted to hospitals. For secondary dengue infections, among children aged 1 to 3 years, the sex ratios

Table 2 Relation between Dengue Antibody Status and Occurrence of DHF-DSS in Prospectively Studied Populations

Study	No. primary infections	No. primary DSS	Primary DSS per 1000	No. secondary infections	No. secondary DSS	Secondary DSS per 1000
Bangkok 1962[a,b]	205	0	0	150	5	33.0
Bangkok, projected to whole city[c]	165,794	12	0.07	125,728	1428	11.4
Koh Samui 1966[d]	26	0	0	83	3	36.0
Koh Samui, projected to whole island[e]	1,900	0	0	2,700	14	5.5

[a]S. B. Halstead, et al. *Yale J. Biol. Med. 42*:311-328, 1970.

[b]S. B. Halstead, et al. *Am. J. Trop. Med. Hyg. 18*:972-983, 1969.

[c]S. B. Halstead, in *Togaviruses* (R. W. Schlesinger, ed.). Academic, New York, 1980.

[d]P. E. Winter, et al. *Am. J. Trop. Med. Hyg. 17*:590-599, 1968.

[e]P. K. Russell, et al. *Am. J. Trop. Med. Hyg. 17*:600-608, 1968.

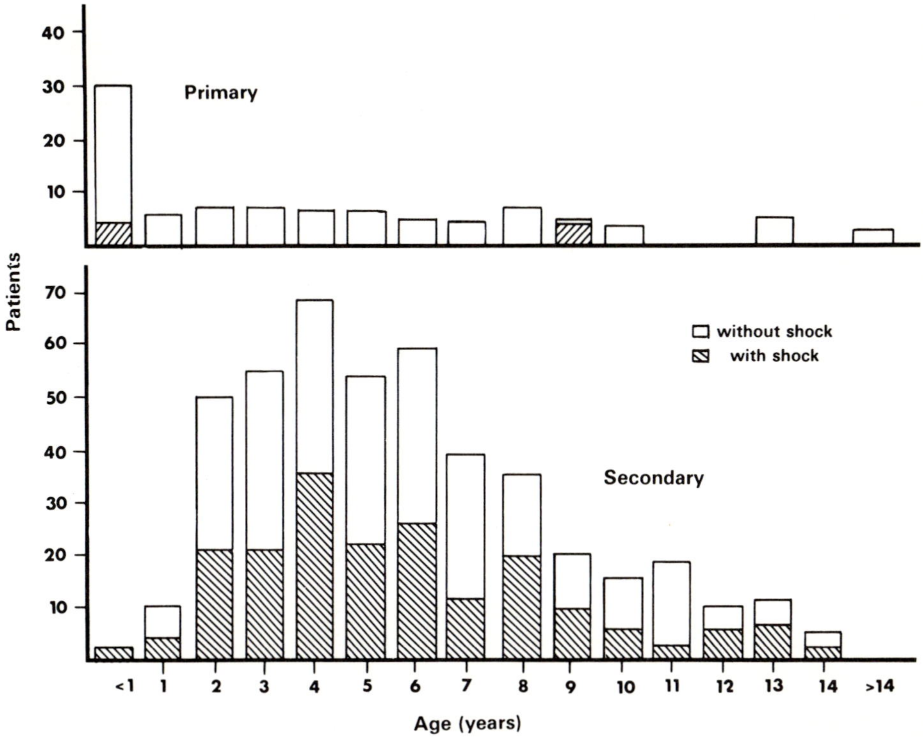

Figure 2 Age distribution of Bangkok children with DHF, Bangkok Children's Hospital, by type of immune response: primary infection, 71 cases; secondary infection, 452 cases. (From S. B. Halstead, S. Nimmannitya, S. N. Cohen, Observations related to pathogenesis of dengue hemorrhagic fever, *Yale J. Biol. Med. 42*:311-328, 1970.)

were nearly equal. However, in 4-year-old and older children there was a striking increase in the admission of girls with DSS compared with boys. The difference was seen in all DHF but was most notable among shock cases. The male-female ratio in a group of 137 shock cases in children $\geq$ 4 years was 1:2.

Nutritional Status

There is a consensus among pediatricians in Southeast Asian countries that children with DHF-DSS are usually very well nourished. Well designed, but unpublished studies comparing DHF cases with age- and sex-matched controls have been undertaken at the Rangoon and the Bangkok Children's Hospitals. In both instances, children with DSS as a group were significantly taller and heavier for age than were normal control children. DHF-DSS is rarely seen in children with clinically apparent protein-calorie malnutrition. In some urban areas, DHF-DSS attack rates have been higher among the middle and upper income groups.

Sequence of Infection

Since four types of dengue virus are known to exist, and DHF-DSS is endemic in areas where three or four types are in continuous circulation, it is natural to wonder whether severe illness during secondary dengue infections results from infections which occur in certain sequences. Unfortunately, the longitudinal epidemiologic studies which might have clarified this point have not been done. Circumstantial evidence relating to the possible impact of infection sequence was obtained in a 3 year study of dengue infections among outpatients and inpatients at the Bangkok Children's Hospital, in foreign nationals resident in Bangkok, and in wild-caught *Aedes aegypti* during 1962-1964. The proportionate distribution of dengue viruses isolated from mosquitoes and from individuals experiencing a primary infection was compared with virus isolates from patients experiencing secondary dengue infections. These distributions are shown in Table 3. It will

Table 3 Distribution of Dengue Viruses Isolated from Susceptible and Dengue-Immune Hosts by Severity of Illness

	Dengue virus type							
	Susceptible				Dengue-immune			
Disease group	1	2	3	4	1	2	3	4
DF or FUO								
outpatients	12	14	15	1	3	11	5	
mosquitoes	6	8	8	2				
DHF, nonshock (hospitalized)	13	14	10		5	22	14	1
DSS (hospitalized)	3				2	15	3	1
Totals	34	36	33	3	10	48	22	2
%	32	34	31	3	12	59	27	2

Source: S. B. Halstead, S. Nimmannitya, S. N. Cohen, Observations related to pathogenesis of dengue hemorrhagic fever, *Yale J. Biol. Med. 42*:311-328, 1970.

be observed that dengue types 1, 2, and 3 viruses were recovered at very nearly equal rates from individuals experiencing their first infection. Dengue type 2 was recovered at high frequency from patients with secondary dengue infections, especially those patients with dengue shock syndrome. Dengue 2 viruses have been recovered at approximately this same frequency from DHF-DSS patients in Bangkok for the entire period 1962-1980. One explanation for this observation is that one or more of the following sequences of infection are pathogenic: dengue 1-dengue 2; D3-D2 or D4-D2. A DHF-DSS outbreak on Koh Samui Island in the Gulf of Siam in 1967 appears to be an exception to this generalization. Only dengue 4 viruses were recovered from mosquitoes collected during this outbreak. Because of the limited geographic scope of virus isolation attempts in comparison with the distribution of cases, it cannot be concluded that dengue 2 transmission did not occur. More recently, large numbers of dengue 3 viruses have been recovered from DHF-DSS and fatal shock cases in Indonesia. Because comparable virus isolations from primary human dengue infections were not reported, estimates cannot be made of denominator data to show whether dengue 3 viruses were associated at higher frequency with DSS in comparison with mild primary dengue infections.

What may be negative observations have been reported from areas of the world with sequential dengue infections where outbreaks of DHF-DSS have not been described. These are summarized in Table 4. The most carefully documented sequential infections are those experienced on islands. For example, in Puerto Rico there were successive outbreaks of dengue 3, dengue 2, and dengue 1 in 1963, 1970, and 1977, respectively. In the Pacific, dengue 3 was endemic on Tahiti from 1963-1969, dengue 2 was introduced in 1971-1973, dengue 1 in 1975, and dengue 4 in 1979. Dengue 2 outbreaks were widespread on the Pacific islands in 1971-1972; many of the same islands experienced dengue 1 epidemics in 1974-1975. In none of these instances did a large outbreak of dengue 2 occur following an earlier experience with a different dengue virus in which there was a relatively *short* interval between outbreaks. This can be contrasted with the 1927-1928 dengue experience in Greece. In 1928, an explosive epidemic occurred which resulted in over 1,000,000 reported cases and 1200 deaths. Neutralization studies on sera collected from Athens residents born in 1927 and 1928 have demonstrated monospecific dengue 1 and 2 antibodies, together with heterospecific antibodies. Dengue 1 and 2 antibodies were distributed by month of birth in a pattern which suggested that an early, large dengue 1 epidemic was succeeded 6 months to 1 year later by a smaller dengue 2 outbreak.

Table 4 Countries Experiencing Sequential Epidemics Due to Different Dengue Viruses without Accompanying Outbreaks of DSS

Country	Date	Virus	Date	Virus	Date	Virus	Date	Virus
Puerto Rico	1963	D3	1969–1977	D2	1977	D1, 2, 3		
Colombia	1971–1972	D2	1975–1979	D3				
Dominican Republic	1972–1973	D2, 3						
Jamaica	1963	D3?	1968–1969	D2, D3	1977	D1		
Tahiti	1963–1969	D3	1971–1973	D2	1975	D1	1979	D4
Fiji	1940s	D1	1971–1973	D2	1974–1975	D1		
Somoa	1940s	D1	1972	D2				
Tonga	1930	D1	1972–1974	D2	1974–1975	D1		

Source: S. B. Halstead, Immunological parameters of togavirus disease syndrome. In *Togaviruses* (R. W. Schlesinger, ed.). Academic, New York, 1980.

Virus Virulence

During the earliest recognized outbreaks of DHF-DSS in Manila and Bangkok, four apparently new viruses were isolated. These were named dengue types 3, 4, 5, and 6. Initially it was believed that these new types were usually pathogenic and were responsible for the shock and hemorrhagic manifestations seen. On further comparison, type 5 dengue proved to be a strain of dengue 2, and type 6, a strain of dengue 1. It is now recognized that dengue types 3 and 4 have been the cause of outbreaks of typical dengue fever not accompanied by DHF-DSS. This provides evidence that not all strains of dengue 3 and 4 are inherently virulent.

Since dengue shock syndrome has been observed during acceptably documented primary infections in a small number of children older than 1 year whose clinical diagnosis was established by experienced pediatricians (Bangkok Children's Hospital), one possible interpretation of this observation is that some dengue strains possess the biologic capability of causing DSS. The spectrum of disease produced in susceptible and immune individuals is diagramatically represented in Figure 3. A very small percentage of primary infections manifest DHF-DSS; disease response during secondary infections is shifted toward increased severity. The increased risk of DHF-DSS in secondary dengue has been studied quantitatively. During the Bangkok epidemic of 1962 it was estimated that a secondary dengue infection conferred a 120-fold greater risk of developing shock syndrome than a primary infection (Table 2).

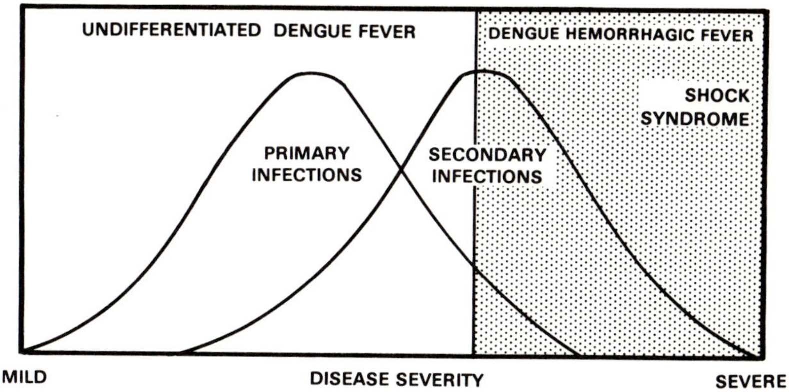

Figure 3 Schematic representation of the "shift" in severity associated with secondary as compared with primary dengue infections.

There is also the possibility that the host *always* determines the severity of dengue infections. In experimental infections of gibbons and monkeys it was demonstrated that a disease process or an unrelated antigenic stimulus may nonspecifically enhance the severity of a dengue infection. This topic is discussed more fully below (see Immune Enhancement Hypothesis).

A number of other accounts of severe disease accompanying "primary" dengue infection have been reported. Most of these are described in adults or are fatal cases of "DSS" in children. The majority of these cases either fail to satisfy the minimum diagnostic criteria of DHF-DSS (no evidence of increased vascular permeability accompanied by thrombocytopenia) or fail to satisfy strict criteria for a primary dengue infection. A child inappropriately given a blood transfusion or who is overhydrated will develop heart failure and may appear to die in "shock." The differentiation of primary and secondary dengue infections in patients with fatal outcome poses an interesting challenge for correct serologic identification. In the typical case of secondary infection DHF-DSS, the patient is hospitalized late in the disease (the mean day of admission is 4 to 5 days after onset of fever). Usually by this time, the patient has HI antibody at a relatively high titer. However, in 8% of a large series of children with DHF-DSS, the acute phase serum contained no detectable HI antibody. However, if tested, these patients do have monospecific neutralizing antibody residual from the first dengue infection. It appears that heterospecific flavivirus group antibody is complexed by viremia to undetectable levels. This conclusion is supported from studies on secondary dengue infections in rhesus monkeys. A *negative phase* secondary response is described for many immunologic systems. There are important consequences of the negative phase immune response. If a patient dies during this phase of antibody response, his or her serum will demonstrate monospecific neutralizing antibody, suggesting a primary infection. However, this antibody *will be of the IgG class*. Only if monospecific IgM antibodies are detected can a fatal case be considered to be undergoing a primary infection. Until the present time not a single fatal DHF-DSS case in a child aged 1 year or older has been documented as a primary dengue infection using acceptable diagnostic criteria.

EXPERIMENTAL STUDIES ON DENGUE

Mention has been made of the almost complete absence of virus and viral antigen from the tissues of fatal DHF-DSS. Because of this problem, dengue pathogenesis must be studied in a model system. The rhesus monkey has provided a useful model of human dengue.

Observations on monkey dengue has led to several fundamental pathogenetic discoveries in humans.

Subhuman primates are highly susceptible to peripheral infection by all types of dengue viruses. In a series of 40 rhesus monkeys infected with dengue 2 (16681 strain), tissues were obtained for virus isolation and fluorescence microscopy. A composite time course of infection was constructed from these data. Following subcutaneous inoculation, virus can be recovered from the inoculation site. While still persisting at the inoculation site, within 24 h it appears in the regional nodes. Virus can also be found 1 day later in other lymph nodes—thoracic, abdominal, inguinal. Virus can be recovered 3 to 5 days after inoculation from the inoculation site: regional and general lymph nodes, bone marrow, spleen, gastrointestinal tract and lung; viral antigen can be visualized in Kupffer cells in the liver. It is during this period that free virus can be recovered from the blood (viremia). Dengue viremia in rhesus monkeys persists for a mean period of 4 days. At the end of the viremic period, virus can be recovered from circulating leukocytes and from skin all over the body. For up to 1 to 3 days *after* viremia has ended, virus can be recovered from multiple tissues.

Three features of dengue in rhesus monkeys are notable.

Virus persists at each site which becomes sequentially infected. The effect of this phenomenon is to produce a cumulative increase in the number of infected cells. Maximum cellular infection is reached just *prior* to the immune elimination of intracellular infection.

Peripheral blood monocytes, infected in vivo, circulate in the blood at a time regularly coordinated with other infection events. Infected monocytes are found in the blood just before or just after the end of the period of viremia. This phenomenon has also been observed in humans, both in a carefully studied accidental laboratory infection with dengue 2 strain 16681 (unpublished observations, this laboratory) and in DHF-DSS cases. Workers in Thailand have recovered dengue viruses from glass-adherent blood leukocytes obtained from large numbers of DHF-DSS patients. Circulating leukocytes are separated by centrifugation on Ficoll-Hypaque, washed free of antibody, and added to dengue-susceptible tissue culture cells. Since the typical DHF-DSS patient is hospitalized at defervescence, the "acute" phase blood specimen is taken at about the end of the viremic period, the time when infected leukocytes are in the blood.

Tissue infection persists after viremia ends. This is an important observation for two reasons. First, it implies that although antibody is capable of neutralizing extracellular virus, early antibody is not efficient at eliminating virus-infected cells. Second, the period in monkeys of the immune elimination of virus-infected cells (postviremia) coincides with the period in humans of maximum disease (postdefervescence).

Cellular Localization of Dengue Infection

In unpublished studies in this laboratory, three dengue 4-immune rhesus monkeys were infected subcutaneously with dengue 2 (16681 strain). The animals were sacrificed 2, 5, and 7 days later. Each animal was injected intravenously with colloidal carbon 1 day prior to sacrifice. Samples of 31 tissues were studied for virus isolation and fluorescent antibody (FA) localization of dengue antigens.

On day 2, virus was isolated from bone marrow and regional lymph nodes; on day 5, from bone marrow, lymph nodes, and gastrointestinal tract. In the day 7 animal, virus was recovered from skin samples from the right forearm, abdomen, and left leg; from liver, duodenum, jejunum, ileum, esophagus, and heart. The animal was viremic on days 4 to 6 but not on the day of autopsy. In the gastrointestinal tract, virus was localized to mononuclear cells which had ample cytoplasm. These were scattered throughout Peyer's patches. In the spleen, FA-positive cells were always localized to paracortical areas rich in carbon-containing macrophages. Small amounts of carbon were found in the cytoplasm of many fluorescing cells. In the liver, scattered cells with bright cytoplasmic fluorescence were seen. These cells had a narrow shape, were elongated, and resembled Kupffer cells. About one-half of FA-positive cells contained one to several carbon particles.

In previously published studies on dengue pathogenesis in rhesus monkeys, dengue antigen was localized to cells with histiocyte morphology in the skin and to isolated mononuclear leukocytes in lymph nodes.

The studies described provided the only information available on the site(s) of dengue infection in experimentally infected primates. The results suggest that dengue viruses replicate in cells of mononuclear phagocyte lineage. Hosts for dengue replication may be one or more macrophage subpopulations. In monkeys, dengue antigen was seldom detected in cells actively phagocytic for carbon particles. The identity of non-carbon-bearing dengue antigen-positive cells in gut lymphoid tissue has not been established.

Enhanced Dengue Infections in Flavivirus-Immune Hosts

In an attempt to recapitulate in monkeys the epidemiologic events antecedent to human DHF-DSS, an unique phenomenon was discovered. More than 100 monkeys were infected in sequence with all four types of dengue viruses given in every possible combination. First and second infections were done at intervals of 2, 6, and 12 weeks and 6 months. Unexpectedly, animals which were immune either to dengue 1, dengue 3, or dengue 4 viruses and then challenged with dengue 2 circulated virus at *higher* mean peak titers than did nonimmune animals which were similarly inoculated with dengue 2 virus using the same dose and route of administration.

Workers at the Walter Reed Army Institute of Research (WRAIR) have inoculated an attenuated strain of dengue 2 (Puerto Rican strain PR159, S-1 clone) into susceptible and yellow fever-vaccinated adult human beings. Differences in seroconversion rates and in the frequency of dengue-like signs were observed. Among 15 yellow fever immunes given dengue 2 vaccine at dosages ranging from 300 to 300,000 plaque-forming units (PFU), there were 10 who seroconverted. By contrast, only 5 of 14 nonimmune subjects similarly inoculated developed neutralizing antibody. Altogether, the WRAIR group has made clinical observations on 9 susceptible adults and 10 yellow fever immunes who seroconverted following dengue 2 vaccination. The frequency of leukopenia and fever was greater in the yellow fever immunes than in the nonimmunes. Although neither of these associations is statistically significant, both sets of data are compatible with an enhancement of dengue 2 vaccine virus infection in yellow fever immunes. These results may be the in vivo counterpart of in vitro experiments described in Antibody-Enhanced Dengue Infection.

Monocytes from Dengue-Immune Primates

Blood monocytes obtained from dengue-immune rhesus monkeys or human beings have been found to be permissive to infection by dengue viruses, in vitro. Mononuclear leukocytes are separated from other cells using the Ficoll-Hypaque density flotation method and then suspended at a concentration of 1 or 1.5×10^6 cells per milliliter in RPMI-1640 medium supplemented with fetal calf serum. This procedure routinely dilutes donor serum to 1:40,000 or greater. Dengue viruses replicate in cultured monocytes when the virus is added at a multiplicity of infection with respect to total mononuclear leukocytes of 0.01 or greater. Monocytes comprise about 10% of blood mononuclear leukocytes.

The receptor which mediates virus infection can be demonstrated to be cell associated; repeated washing of cells to dilutions of original serum in excess of 1:10,000,000 does not ablate dengue virus permissiveness. Treatment of cells with trypsin and anti-Ig reduced or destroyed permissiveness. These experiments suggest the existence of a cytophilic antibody which is firmly adherent to monocytes. Since the addition of enhancing antibody (see below for discussion of enhancing antibody) to dengue-immune monocytes produces an additive effect, it may be surmised that there are separate monocyte populations, each infected by a different mechanism. The association of antibody and cell, if this occurs, must take place only in vivo, since the incubation of anti-dengue antibodies with monocytes in vitro does not result in irreversible attachment of infection-enhancing antibody to cells. Further work is needed to identify the receptor for dengue virus on human monocytes. It is possible that "natural" antibody could be cytophilically associated with monocytes and serve as a receptor for virus in nonimmune individuals.

Antibody-Enhanced Dengue Infection

In Vitro Infection

When it became apparent that immune status regulated the infection of mononuclear leukocytes in vitro, it was logical to inquire if this phenomenon was mediated by antibody. It has now been established in several laboratories that the addition of heterotypic dengue antibody, nondengue flavivirus antibody, or relatively high dilutions of homotypic dengue antibody to cultures containing dengue virus and monocytes increases the number of monocytes infected with virus and the amount of infectious virus produced. This phenomenon has been extended to Murray Valley encephalitis virus in chicken monocytes and to yellow fever and West Nile viruses in a continuous mouse macrophage line ($P388D_1$ cells) and a human histiocytoma line (U937 cells). Dengue 2 virus also exhibits antibody-enhanced infection in these stable mouse and human macrophage cell lines.

The initial stages of the infection of monocytes by dengue 2 virus have been well studied. Dengue infection is dependent upon the formation of immune complexes by one or more subtypes of IgG. The immune complex attaches to monocyte Fc receptors as is evidenced by the requirement for a Fc terminus [$F(ab)_2$ immune complexes do not enhance] and the ability of irrelevant immune complexes and a monoclonal anti-Fc receptor to block infection enhancement. The ability of $F(ab)_2$ immune complexes to enhance infection of monocytes can be restored by forming a new immune complex with anti-human IgG. Dengue virus-antibody complexes are irreversibly attached to monocytes within 10 min and internalized 20 to 50 min later.

It is suspected that in suspension cultures of peripheral blood mononuclear leukocytes only a small subpopulation of monocytes is permissive to infection by dengue immune complexes. This is based upon evidence that usually not more than about 1% of monocytes is infected. Further, dengue virus permissiveness decreases if cells are held in culture for 2 days or more before addition of immune complexes and that bone marrow suspensions readily demonstrate infection enhancement, replicating virus to high titers. These observations have been interpreted to suggest that immature monocytes possessing the property of immune phagocytosis but not actively producing lysosomal enzymes may be particularly susceptible to dengue infection. Promonocytes comprise about 1% of the peripheral blood monocyte population.

Other than bone marrow cells, little is known of the ability of tissue mononuclear cells (e.g., histiocytes, macrophages, Kupffer cells, reticulum cells), to support antibody-mediated dengue replication in vitro. The absence of data should not be interpreted to mean that such cells are not infected in vivo. Quite the opposite is the case.

In Vivo Infection

Since antibody of the appropriate specificity can enhance dengue virus infection of human and rhesus monkey monocytes in vitro might the same phenomenon operate in vivo? Infection enhancement by passive transfer of antibody fits the epidemiologic conditions described for infants less than 1-year-old who develop DHF-DSS. This question was studied experimentally in five pairs of susceptible rhesus monkeys inoculated intravenously either with normal or dengue-immune human cord blood serum effecting a final dilution of 1:300. The immune serum pool had a dengue 2 plaque reduction neutralization titer of 1:140 and a human monocyte infection enhancement titer of greater than 1:1,000,000. Animals were infected 15 min after inoculation of serum. In each of the comparison pairs, animals receiving dengue antibody had higher viremia levels than their controls. Cumulative daily viremia titers were 2.7- to 51.4-fold greater in animals receiving dengue-immune cord blood serum compared with nonimmune cord blood serum.

Enhancing Antibody

Homotypic dengue antibody or antibodies from multiple infected humans or monkeys neutralize dengue viruses at low dilutions, but at dilutions at or above the neutralization end point, infection enhancement occurs. Dengue antibodies raised to heterotypic dengue viruses or other flaviviruses often exhibit heterospecific neutralization at low serum dilutions; at higher dilutions these sera enhance (Fig. 4). Serum enhancement titers for dengue 2 (16681 strain) often are 100- to 10,000-fold higher than hemagglutination inhibition or neutralizing titers against the homologous virus.

When representative flavivirus sera were tested against a battery of flaviviruses using the neutralization and HI tests and the same sera studied for dengue 2 infection enhancement, those sera demonstrating the strongest cross-reactivity in the HI test and the greatest enhancing activity. Data summarized in Table 5 clearly relate the enhancement phenomenon to antibody interactions with flavivirus group determinants. Among the dengue viruses, neutralizing and protecting antibodies are type specific, an observation which implies the existence of two populations of antibodies—type specific which neutralize and group specific which enhance. Monoclonal antibodies raised in mire do demonstrate the existence of type-specific antibodies which neutralize but do not enhance and group-specific antibodies which enhance, but do not neutralize.

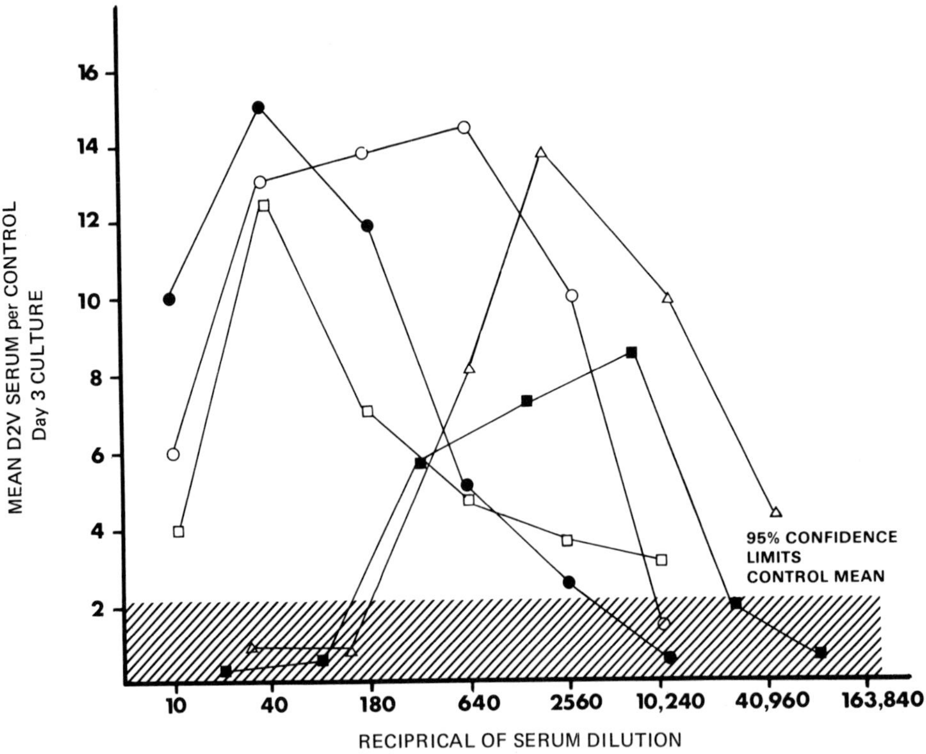

Figure 4 Titration in human peripheral blood leukocyte cultures of dengue 2 infection-enhancing activity of anti-dengue 1 (open squares), anti-dengue 2 (solid squares), anti-dengue 3 (open circles), anti-dengue 4 (closed circles), and multitypic anti-dengue 1 to 4 (triangles). Values above the cross-hatching represent significant enhancement of dengue virus replication at the indicated dilution of serum. (From S. B. Halstead and E. J. O'Rourke, Dengue viruses and mononuclear phagocytes, *J. Exp. Med. 146*:201-207, 1977.)

Antibody-mediated infection enhancement of dengue viruses is largely the property of IgG immunoglobulins. Enhancing anti-dengue IgG appears in the blood relatively early following infection of human or monkey, but can only be detected by separating the IgM and IgG immunoglobulins. When mixed together, the relatively type-specific neutralizing IgM blocks the immune enhancement phenomenon. Anti-dengue IgG does not require complement to enhance. Weak enhancement has been demonstrated by adding complement to serum fractions

Table 5 Dengue 2 Virus Infection Enhancement in Human Monocytes by *Flavivirus* Antisera

Subgroup		Homologous HI titer	D2V HI titer	No. of reacting Ag.	D2V Enh titer	D2V PRNT
Tick-borne	Langat	160	20	12	320	10
	Louping Ill	ND	ND	ND	320	20
	Entebbe bat	160	20	8	80	10
Bat	Bukalasa bat	20	20	5	320	10
	Modoc	80	160	*40*	*10240*	20
Mosquito	Japanese enceph.	80	40	*36*	*20480*	20
	West Nile	80	ND	*35*	*2560*	20
	St. Louis enceph.	ND	ND		320	20
	Usutu	160	ND	*32*	*1280*	20
	Kokobera	160	20	5	40	10
	Alfuy				80	20
Spondweni	Spondweni	160	20	11	10	ND
	Zika	80	20	10	80	20
	Chuku				1280	20
Israel Turkey	Israel Turkey	160	20	9	640	10
	Ntaya	40	20	17	*1280*	20
	Tembusu				1280	20
Banzi	Uganda S	640	40	*37*	640	10
	Edge Hill	1280	20	4	320	20
Dengue	Dengue 1	1280	20	ND	2560	20
	2	120	160	ND	10240	160
	3	20	20	ND	2560	20
	4	80	20	ND	40	10
Ungrouped	Bussuquara	20	20	ND	80	20

Source: S. B. Halstead, J. S. Porterfield, E. J. O'Rourke, Enhancement of dengue virus infection, *Am. J. Trop. Med. Hyg. 29*:638–642, 1980.

containing anti-dengue IgM. Finally, human and monkey cord blood serum from infants born to dengue-immune Southeast Asian mothers contains enhancing antibody, often at very high titers. Since human monocytes possess IgG_1 and IgG_3 receptors and these two subclasses of IgG are actively transported across the human placenta, it can be surmised that enhancing activity resides in one or both of these IgG subclasses. There are no reports of definitive studies on the IgG subtype of enhancing antibody. The role of IgM is less well defined, but presumably IgM antibodies directed at determinants which do not mediate neutralization can enhance by attaching to monocyte plasma membranes via the C3 receptor.

PATHOGENESIS OF DHF-DSS—THE IMMUNE ENHANCEMENT HYPOTHESIS

Any hypothesis which seeks to explain the pathogenesis of DHF-DSS must reconcile various clinical, pathologic, and epidemiologic observations:

1. DHF-DSS is an acute vascular permeability syndrome accompanied by thrombocytopenia.
2. The severity of DHF-DSS is directly correlated with the degree of activation of the complement and blood clotting systems.
3. Increased vascular permeability occurs swiftly at about the time of defervescence.
4. At death, virus and viral antigens are usually absent; both virus and antibody concentrations are lower in fatal cases than in DSS patients who survive.
5. Autopsy findings include evidence of vascular permeability, but no endothelial necrosis or generalized thrombosis; there is marked lymphocyte necrosis in T-dependent areas of spleen, lymph nodes, and thymus.
6. DHF-DSS occurs at relatively high frequency during primary dengue infections in infants less than 1 year of age.
7. DHF-DSS occurs at relatively high frequency during secondary dengue infections.
8. DSS occurs at higher frequencies in females than in males 4 years and older, and in well-nourished than in poorly nourished children.

The major features of the immune enhancement hypothesis are diagrammed in Figure 5.

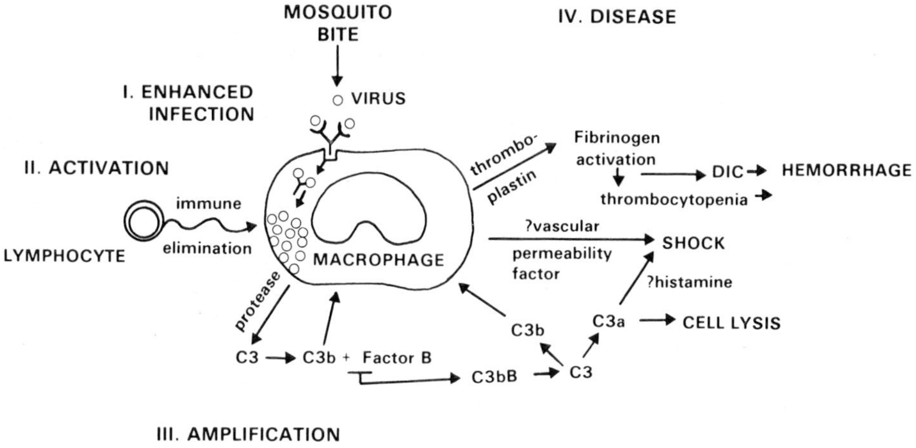

Figure 5 Schematic representation of the immune enhancement hypothesis of the pathogenesis of DHF-DSS. I. Nonneutralizing antibody, whether free or cytophilically bound, mediates dengue virus infection of mononuclear phagocytes. II. Virus replicates and marks cell as non-self, eliciting an immune response which initially activates cell to II. release of enzymes, which include proteases and thromboplastin, which interact with IV. complement and blood clotting systems and may release a vascular permeability factor resulting in the physiologic changes which characterize DHF-DSS.

Afferent Mechanisms—Control of Infection by Antibody

Nonneutralizing antibody, whether passively or actively acquired, facilitates the attachment and internalization of dengue virions in cells of mononuclear phagocyte lineage (i.e., reticuloendothelial system). This provides initial regulation of disease outcome. Antibody serves as a specific receptor. Unlike cell receptors which are numerically fixed for individuals in a species, the size of the enhancing antibody population can vary widely, which may result in a wide variation in the total number of infected cells. This segment of the general hypothesis is supported by experimental evidence.

There are two axioms: (1) The number of enhancing antibody molecules is directly related to the number of cells which become infected. (2) The more cells infected, the more severe the illness.

Activation of Infected Mononuclear Phagocytes

The mechanism of elimination of dengue-infected cells has not been established. In our laboratory, killer leukocytes which attack autologous dengue-infected monocytes appear in the blood 5 to 7 days after dengue infection in rhesus monkeys. Assuming these cells to be T lymphocytes, it is predicted that early interactions between dengue-infected mononuclear phagocytes and T lymphocytes may result in activation of the cells. Such a phenomenon might be related to the observation that dengue virion envelope antigens are not easily demonstrated on the surface of dengue-infected cells. This may make cell-cell killing a relatively inefficient process, allowing monocytes to "read" lymphokines as activation stimuli. Whatever the mechanism, the release of vascular permeability mediators coincides with the time and completeness of elimination of intracellular dengue infection.

Amplification

Mononuclear phagocytes are uniquely interconnected with the complement system. Activated macrophages release lysosomal enzymes, which cleave C3 to C3a and C3b. Activated C3b with properdin factor B functions as a protease, cleaving further C3 to C3b. C3b acts on macrophages to promote the further release of lysosomal enzymes. Thus, through positive feedback mechanisms, macrophages are potentially powerful activators of complement via the alternative pathway.

Effector Mechanism

The same stimuli which result in cleavage of C3 by macrophage products can also elicit the release of thromboplastin. The activation of these two effector systems by macrophage products would explain why the C3 consumption and fibrinogen split products vary cotemporally and directly with the severity of disease (i.e., number of cells infected).

The mechanism of vascular permeability in DSS is still an unsolved mystery. It has been suggested that histamine mediates vascular permeability in DSS. This is the simplest hypothesis, since histamine release is the expected byproduct of the action of C3a anaphylatoxin on mast cells and basophils and, thus, should be related to complement activation. However, the cumulative evidence for histamine-mediated vascular permeability is still weak. There is but a single report of increased histamine excretion in DHF-DSS cases. Dengue shock is notoriously resistant to antihistamines and steroids; shock is never accompanied by histamine-mediated stigmata, i.e., urticaria or erythema nodosum. It is possible that an as yet unidentified macrophage-generated mediator is principally responsible for shock.

C3a is a powerful cytolysin. An interesting possibility is that the marked lymphocytolysis observed in spleen, lymph nodes, and thymus and focal destruction of hepatocytes may be due to local release of C3a by activated macrophages.

THE IMMUNE ENHANCEMENT HYPOTHESIS ACCOMMODATES ALL IMPORTANT OBSERVATIONS ON DHF-DSS

Figure 6 shows that the immune enhancement hypothesis accommodates each of the major observations on DHF-DSS.

Efferent Mechanisms

Passive or actively acquired group-reactive nonneutralizing antibody (observations 5 and 6) produce the same effect—enhanced infection. Enhanced infection implies enhanced viremia and antigenemia which can result in increased formation of immune complexes with removal of both virus *and* antibody from circulation (observation 3). Enhanced antigenemia during *secondary* dengue infections may contribute to the activation of complement via the classic pathway (observation 1). Enhanced infection of mononuclear phagocytes implies the possibility of increases in cellular products (observations 1 and 4).

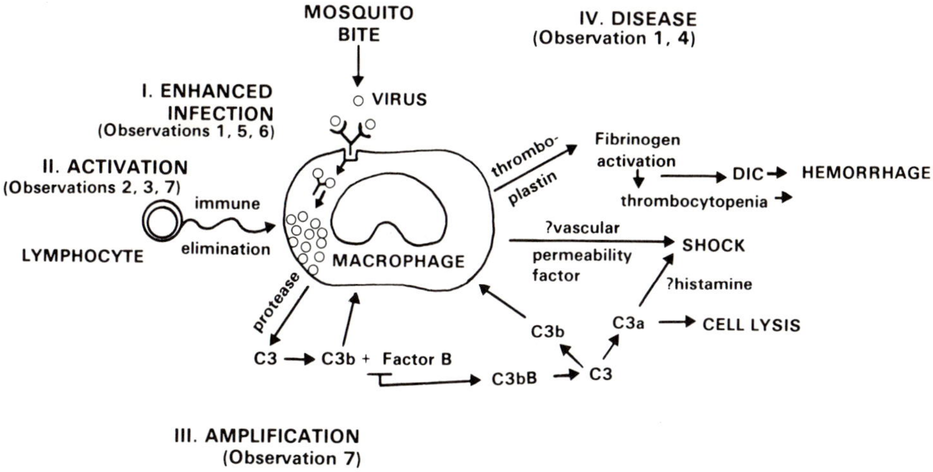

Figure 6 Segments in the immune enhancement hypothesis which correspond to various clinical or epidemiologic observations on DHF-DSS. Some stages in the pathogenetic cycle may be promoted or inhibited with resultant effect on disease outcome. For example, malnutrition may reduce the mononuclear phagocyte activation signal, enzyme production or release and thereby reduce intensity of effector phenomena (disease).

Activation

With respect to their ability to cope with infection, females are considered to be immunologically more competent than males (observation 7). This could be expressed via activation of mononuclear phagocytes or (see below) the efferent limb of cellular immune responses. Protein-calorie malnutrition (PCM) is known to depress the effector limb of CMI. In the immune enhancement hypothesis, immune competence translates into increased efferent phenomena: vascular permeability, and activation of the complement and blood clotting systems, and vice versa. Thus, femaleness would be expected to enhance the severity of DHF, while PCM would be expected to decrease the severity of dengue shock syndrome. The other important attribute of immune competence is success in eliminating cells marked non-self. Using conventional virus isolation systems, the usual patient who dies or has DHF-DSS succeeded in making himself virologically sterile (observation 3).

Amplification and Effector Functions

The role of sex-related immunocompetence and the immunosuppressive effect of protein-calorie malnutrition on effector functions of mononuclear phagocytes have not been studied exhaustively. There is some evidence that these two phenomena could act at this level (observation 7).

It can be seen that the immune enhancement hypothesis provides a logical explanation of the eight essential features of DHF-DSS. There are a number of other observations on the behavior of dengue virus-human interactions which must be explained. Research approaches which might reasonably contribute to an understanding of these phenomena are described, together with appropriate research questions.

WHY DHF-DSS OCCURS DURING SOME PRIMARY DENGUE INFECTIONS

A small number of well-authenticated mild DSS cases have been described during primary dengue infections in children aged 1 year or older.

There are several plausible host- or virus-related phenomena which might result in enhanced dengue infections by a mechanism other than antibody. In reflecting on the implications of the immune enhancement hypothesis it is self-evident that as opposed to parenchymal cells, which are fixed in number, mononuclear phagocytes belong to a family of cells which can expand or contract, and which can be activated or functionally blocked by various stimuli.

If the promonocyte is an important host of dengue infection, any stimulus which increases production of mononuclear phagocytes should increase the absolute numbers of promonocytes. This could include acute or chronic infections or lymphoproliferative disorders. Enhanced dengue viremia has been observed in experimentally infected gibbons which had concurrent lymphomatous leukemia. Enhanced viremias have been produced in gibbons and rhesus monkeys which were inoculated with pertussis vaccine (gibbons and rhesus) or *Corynebacterium parvum* (rhesus) just prior to dengue infection.

Engulfment of dengue viruses by mononuclear phagocytes may increase without a corresponding expansion of the cell population. In mice, the administration of several steroid hormones increased phagocytosis by macrophages. In unpublished studies in this laboratory, progesterone and other steroid hormones markedly increased dengue 2 infection in human monocytes in vitro. If steroid hormones in the human female more effectively stimulated phagocytosis than in the male, this might help to explain the sex differential in the occurrence of dengue shock syndrome.

Finally, a dengue virus which has enhanced replicative or survival properties might also produce enhanced infections. If such mutant strains exist, it may be possible to assess their "virulence" by comparing the ability of different dengue strains to replicate in human monocytes in vitro.

Once the kinetics of infection are altered, target cell activation, amplification, and effector mechanisms would proceed as usual. The simplest explanation for the smooth gradation in severity of all dengue infections, from mild to severe, is that infections differ merely in the number of cells serving as replicative hosts for virus.

WHY DHF-DSS DOES NOT OCCUR DURING ALL SECONDARY DENGUE INFECTIONS

This question has been troublesome to many workers in the dengue field. There is reasonable evidence in Thailand that 1 to 3% of all secondary dengue infections produce shock syndrome. For each DSS case hospitalized, there is usually an additional patient with DHF (nonshock) admitted to the hospital. This provides the estimate that secondary dengue infections result in 2 to 6% of children with a serious disease, clearly a small minority.

There are several possible mechanisms which might explain the relative infrequency of severe illness during secondary dengue infections.

Heterotypic Neutralizing Antibody

The simplest hypothesis is that infections by most dengue viruses raise heterotypic neutralizing antibodies, which at low dilution are effective in modulating the enhancement phenomenon. Only the rare individual has *no* heterotypic neutralizing antibody. Such an individual is "at risk" of developing DHF-DSS. The effect of heterotypic neutralizing antibody is observed frequently and is illustrated in Figure 6. The absence or depression of dengue infection in monocytes treated with low dilutions of dengue 1, 3, and 4 antiserum exemplifies heterotypic neutralization.

Sequence of Infection

Epidemiologic studies in Bangkok for the period 1962-1980 have demonstrated a remarkably stable predominance in the recovery of dengue 2 viruses from DSS cases. This phenomenon has been reviewed above. One possible explanation is that one or more of the following sequences of dengue infection may be more pathogenic than infections acquired in other sequences: 1-2, 3-2, or 4-2. Recent recoveries of dengue 3 virus from DSS cases in Indonesia, if verified, may mean that there are other pathogenic sequences. Both the phenomenon of sequence and heterotypic neutralizing antibody may be related to the distribution of antigenic determinants on infecting pairs of viruses (see below).

Interval between Infections

A computer-assisted analysis of age-specific hospitalization rates for DHF-DSS in Bangkok has demonstrated a requirement for an upper limit on the interval between the first and second infection which results in DHF-DSS. This was estimated at 5 years. There is no obvious explanation for this phenomenon in the present formulation of the immune enhancement phenomenon. Nothing is known as yet about the long-term kinetics of enhancing antibodies in human populations. An observation of possible relevance is that in dengue-immune humans and rhesus monkeys, permissive monocytes eventually disappear from the blood. Are these cells evidence of the existence of a permissive population more important than circulating enhancing antibodies in the immune-enhanced infections? There are no answers to these questions.

It is rather interesting to note that many sequential dengue epidemics in the Caribbean and the Pacific Islands which have not resulted in outbreaks of DHF-DSS have occurred at intervals in excess of 5 years. With the exception of Southeast Asia, where three or four dengue types are endemic, there has been no sizable population in which sequential dengue outbreaks have been documented to occur at an interval of less than 5 years *and* in which type 2 was the last in the sequence. Dengue epidemiology in South India remains enigmatic. In the 1960s, all four dengue viruses were recovered from infected humans in Vellore near Madras. No DHF-DSS outbreaks have ever been reported.

Structure-Function Relationships

Little is known today about the molecular complexity of flavivirus group determinants. Cross comparisons within the flavivirus group has demonstrated that antisera raised to different members of the group inhibit hemagglutination in patterns unique to different viruses. This suggests that there exists a mosaic of flavivirus group determinants. Hits between virus and antibody should be a function of the concentration of antibody molecules in extracellular fluids. Hits should increase if there are several group determinants available to interact with complementary antibodies. It follows that pairs of infecting dengue viruses which share a large number of group determinants should elicit more enhancing antibodies than infecting pairs which share a single common determinant. Thus, the existence in nature of dengue viruses which share flavivirus group determinants extensively may lead to severely enhanced secondary dengue infections. Few shared structural relationships might produce secondary dengue infections of relatively mild nature. Might the accelerated endemic circulation of four dengue viruses in Southeast Asia have resulted in enough recombination to produce increased structural relatedness?

DISCUSSION

This chapter began by asking the reader to accept the notion that the pathogenetic mechanisms in dengue hemorrhagic fever represented a fourth type of immunopathology. The biologic implications of the experimental work summarized can also be viewed in a broader context. Many pathogeneticists would agree that macrophages are the most important initial determinant of virulence in systemic virus infections. Viruses appear to encounter phagocytic cells soon after parenteral invasion. Survival or successful replication is essential if the infection is to continue and if virus ultimately is to reach target cells which produce the characteristic infection syndrome. Genetic and maturational differences in susceptibility of hosts have often been related to the ability of macrophages to destroy virus or serve as replicative hosts.

The enhanced infection of cells by nonneutralizing antibody in the dengue system has drawn attention to the possibility that either antibody, or changes in the function or size of the mononuclear phagocyte population, may regulate the severity of many systemic virus infections. An enduring enigma in many viral diseases is the relatively high ratio of inapparent infections to overt cases. Since inapparent infection appears to be the rule, perhaps some low-frequency phenomenon intervenes to enhance infection in the rare individual who contracts overt disease. Might the "spreading" effect versus the "containment" role of the mononuclear phagocyte system be this low-frequency factor?

Whatever the general implications of immune enhancement, the hypothesis leads to several specific predictions.

First, it will be noted that the effect of enhancing antibody postulated for dengue is to increase infection of mononuclear phagocytes. In situations in which viruses efficiently infect these cells without the assistance of antibody *and* infected mononuclear phagocytes are subsequently activated, an acute vascular permeability syndrome with hemorrhagic manifestations may result. Remarkably, most viral hemorrhagic fevers of humans produce both hemorrhage and a shock syndrome. This group includes yellow fever, Argentine and Bolivian HF, Congo-Crimean and Omsk HF, Kyasanur Forest Disease and the African HF, Lassa, Marburg, and Ebola. Pathologic findings in each of these diseases are remarkably similar. The arenaviruses are notable for their marked tropism for macrophages in experimentally infected animals. The arenaviruses generally do not produce lytic infections in cells, and it is known that T lymphocytes are central to immune elimination and disease production. Might effector systems in all the viral hemorrhagic fevers be similar to those of DHF?

Second, it will be apparent that there are two prerequisites to immune infection enhancement: nonneutralizing antibody and a virus capable of replicating in mononuclear phagocytes. The number of naturally occurring immunologically enhanced infections is unknown, but it will be recalled that the administration of the formalin-killed measles, respiratory syncytial virus, parainfluenza, and mycoplasma vaccines resulted in short-lived protection followed by a return to a susceptible state. Formalin treatment is known to destroy antigenic determinants essential to neutralization; hence, nonneutralizing antibodies are raised. When wild-type virus infections have occurred they produce disease which was markedly more severe than normal. Although explanations of these "atypical" syndromes abound, no serious attempt has been made to assess the possibility that they are examples of the immune enhancement phenomenon.

Finally, the infection-enhancing effect of antibody may be an evolutionary force which selects for antigenic diversity. Limited studies in our laboratory suggest that Murray Valley encephalitis (MVE) virus infection in chickens may be centered in mononuclear phagocytes: MVE may be an avian dengue virus. If so, there would be a survival advantage for multiple flaviviruses through the immune infection enhancement mechanism. More than 60 flaviviruses are recognized today; many of them occupy overlapping ecological niches. Antibody has been studied for its effect on deleting antigenic determinants, but no one has yet examined the possibility that antibody may support the survival of envelope protein mutants.

Whether any or all of these speculations bear fruit, it seems clear that progress made in understanding the interactions between dengue viruses, antibody, and mononuclear phagocytes should begin to unveil the secrets of pathogenesis at a molecular level.

SUGGESTED READINGS

Halstead, S. B. Immunological parameters of togavirus syndromes. In *The Togaviruses: Biology, Structure, Replication.* (R. W. Schlesinger, ed.). Academic, New York, 1980.

Halstead, S. B. Dengue haemorrhagic fever—a public health problem and a field for research. *Bull. WHO 58*:1-21, 1980.

Halstead, S. B., and E. J. O'Rourke. Dengue viruses and mononuclear phagocytes. I. Infection enhancement by non-neutralizing antibody. *J. Exp. Med. 146*:201-217, 1977.

Halstead, S. B., E. J. O'Rourke, and A. C. Allison. Dengue viruses and mononuclear phagocytes. II. Identity of blood and tissue leukocytes supporting in vitro infection. *J. Exp. Med. 146*:218-299, 1977.

Halstead, S. B., and others. Observations related to pathogenesis of dengue hemorrhagic fever. I-VI. *Yale J. Biol. Med. 42*:261-362, 1970.

Nimmannitya, S., S. B. Halstead, and others. Dengue and chikungunya virus infection in man in Thailand, 1962-1964. I-V. *Am. J. Trop. Med. Hyg. 18*:954-1033, 1969.

Russell, P. K., and W. E. Brandt. Immunopathologic processes and viral antigens associated with sequential dengue virus infection. *Persistent Virus Infections: Perspectives in Virology VIII*:263-277, 2973.

Russell, P. K., T. M. Yuill, A. Nisalak, S. Udomsakdi, D. J. Gould, and P. E. Winter. An insular outbreak of dengue hemorrhagic fever. II. Virologic and serologic studies. *Am. J. Trop. Med. Hyg. 17*:600-608, 1968.

Schlesinger, R. W. *Dengue Viruses.* Virology Monographs. Springer-Verlag, New York, 1977.

Technical Guides for Diagnosis, Treatment, Surveillance, Prevention and Control of Dengue Haemorrhagic Fever. World Health Organization. South East Asia Regional Office, New Delhi, 1980.

Winter, P. E., T. M. Yuill, S. Udomsakdi, D. Gould, S. Nautapanich, and P. K. Russell. An insular outbreak of dengue hemorrhagic fever. I. Epidemiologic Observations. *Am. J. Trop. Med. Hyg. 17*:590-599, 1968.

chapter 3

Interferon

A Molecule for all Seasons

ERIK D.A. DE CLERCQ

Rega Institute for Medical Research
Katholieke Universiteit Leuven
Leuven, Belgium

Since its discovery in 1957 interferon has attracted the attention of a wide array of investigators working in such different disciplines as virology, immunology, biochemistry, organic chemistry, molecular biology, pharmacology, and clinical medicine. While interferon was originally discovered as an antiviral agent, it now appears that its antiviral properties are only incidental to its other numerous biological activities, which include, among others, modulatory effects on the immune response and inhibitory effects on cell division. Also, interferon is not a single substance but rather a family of molecules which can all be defined as glycoproteins. There are several forms of interferon, and these can be most adequately be divided into type I ("classic") interferon and type II ("immune") interferon. Within the class of type I interferon, several variants exist which differ from one another in molecular structure (i.e., amino acid sequence, carbohydrate moiety), physical properties, and antigenicity. The same might be true for type II interferon. The type of interferon produced is not always predictable and depends on both the cell source and the nature of the interferon inducer. There are in fact numerous inducers of interferon; these include viruses, microbial components (i.e., endotoxins), synthetic double-stranded polyribonucleotides, low molecular weight chemicals (i.e., tilorone), lymphocyte mitogens, and specific antigens. Thus, whereas interferon could originally be described as a virus-induced cellular product capable of protecting cells against virus replication, this concept has now to be broadened to accommodate (1) the different agents that have been found to induce interferon,

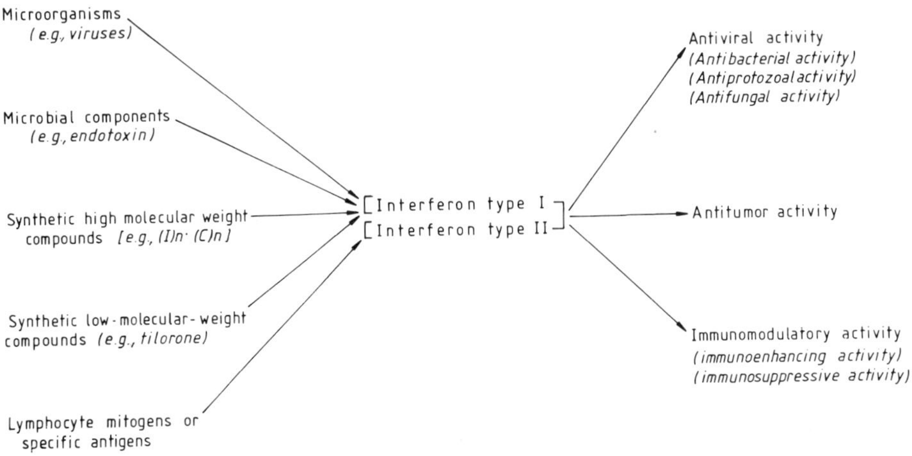

Figure 1 The interferon system.

(2) the different types of interferon produced in response to these inducers, and (3) the different biologic effects exerted by these interferons (Fig. 1). In this chapter I will review the various aspects of the interferon system, and, more specifically, address the following questions:

1. What is interferon and what does it look like?
2. How can interferon be purified?
3. Which are the interferon inducers?
4. How is interferon induced?
5. How can interferon be produced in mass quantities?
6. How does interferon act at the molecular level?
7. How does interferon act at the cellular level?
8. In which experimental situations has interferon proven to be efficacious?
9. In which clinical situations has interferon proven to be efficacious?
10. Are there problems which limit the clinical usefulness of interferon?

Obviously, many more questions could be asked, but those raised cover fairly well the recent trends in interferon research and the advances that have been made during the past few years. I have not attempted to reference any material in the text; instead, a rather extensive reading list is provided at the end, which may be consulted as an up-to-date account on the most recent developments in the interferon field.

STRUCTURE OF INTERFERON

Type I and type II interferon are fundamentally different. Whereas type I interferon can be induced by a great variety of inducers (i.e., viruses, double-stranded RNAs) in a great variety of cells (including T and B lymphocytes), type II interferon is produced only by T lymphocytes in response to T-cell mitogens or specific antigens. Type II interferon can be distinguished from type I interferon by a number of criteria, such as acid stability, antigenicity, and antiviral activity on heterologous cells. Type I interferon is stable at pH 2, whereas type II interferon is not. Type II interferon is not neutralized by antisera to type I interferon, whereas type I interferon is not neutralized by antisera to type II interferon. Type II interferons, whether human or murine, are more species specific in their antiviral activity than type I interferons: i.e., human leukocyte interferon has been found to be highly active on various cell species, including cat, rabbit, and bovine cells, whereas human "immune" interferon has not been found active on these heterologous cell systems. Similarly, mouse type I interferon is significantly active on guinea pig cells, whereas mouse type II interferon is not.

Both the immunosuppressive and immunoenhancing activities are expressed to a much greater extent by type II interferon than they are by type I interferon. Type II interferon would also be a far more potent anticellular agent than type I interferon. This implies that type II interferon is actually a weaker antiviral agent (since the immunomodulatory and anticellular activities are quantitated in terms of antiviral units), and, indeed, cells treated with type II interferon acquire their antiviral resistance much more slowly than cells treated with type I interferon. Thus, while type I interferon could essentially be considered an antiviral agent, type II interferon primarily acts as a cell regulator. Its antiviral activity may be secondary to its general cell-regulatory activity.

Neither type I interferon nor type II interferon corresponds to a single molecular species. For both interferon types there exist multiple molecular forms. For example, for human type I interferon, one mentions generally three varieties of interferon, designated human leukocyte (Le) interferon, human lymphoblastoid (Ly) interferon, and human fibroblast (fibroblastoid; F) interferon, according to the cell types they originate from. Native Le interferon can be resolved into (at least) two protein components, one with a molecular weight of about 16,000 (15,000 to 18,000) and one with a molecular weight of 21,000 (Table 1). These two fractions would occur at an approximate ratio of 7:3. For human F interferon, only one component (with a molecular weight of about 20,000) has so far been identified. Human F interferon produced by fibroblast cells may be identical to the human F

Table 1 Molecular Weights of Native Interferons

Class of interferon	Number of major interferon species	Molecular weight of major interferon species[a]		
Type I interferon				
Human leukocyte (Le) interferon (Hu IFN-α) [human lymphoblastoid (Ly) interferon]	2		15,000-18,000 (16,000) (18,500)	21,000
Human fibroblast (F) interferon (Hu IFN-β) [human fibroblastoid interferon)	1		20,000 (22,000) (19,000)	
Mouse interferon (Mu IFN-α, β)	2		22,000 (24,000)	35,000 (38,000) (40,000)
	3	20,000	26,000	33,000
Type II interferon				
Human immune interferon (Hu IFN-γ)	2		40-46,000	65-70,000
Mouse immune interferon (Mu IFN-γ)	2		40,000	70-90,000 (90,000)

[a]The differences in size between the different interferon species within a given class may, at least in part, be due to unequal glycosylation.

interferon produced by fibroblastoid (aneuploid fibroblast) cell lines. Human Ly interferon, however, would be dissimilar from human Le interferon. Mouse interferon covers a broad molecular weight range (20,000 to 40,000), although it is composed primarily of two major species, a larger one (representing 80 to 85% of the total interferon activity) with a molecular weight of 35,000 (33,000, 38,000, or 40,000) and a smaller one with a molecular weight of 22,000 (24,000 or 26,000) (Table 1). In some mouse inferferon preparations a third species (with a molecular weight of 20,000) has been detected. Unlike human type I interferons, murine type I interferons are not classified as Le (Ly) or F interferon. All major murine type I interferon species may be "F-like." A minor murine type I interferon component, accounting for no more than 1 to 2% of the total interferon activity, behaves antigenically like human Le interferon. This component may represent a "human Le-like" mouse interferon. The type II interferons, both human and murine, can also be resolved into two distinct molecular species, a smaller one (representing 80 to 90% of the total interferon activity) with an apparent molecular weight of about 40,000, and a larger one with an apparent molecular weight of about 70,000 (Table 1).

In contrast with human and murine type II interferons, which have so far not been obtained in pure form, various human type I interferons, Le, Ly, and F, and murine type I interferons have been purified to homogeneity. For all these interferons the amino acid composition has been determined, as has been the amino acid sequence of the amino terminus. This direct amino acid sequence analysis revealed a partial homology in amino terminal sequence between the 18,500 dalton component of human Ly interferon and the 20,000 dalton component of mouse interferon and between human F interferon and the 26,000 dalton and 35,000 dalton species of mouse interferon.

However, the study of the primary structure of interferon has recently been revolutionized by the successful cloning of interferon genes in bacterial systems. This cloning has been achieved with human F interferon (Hu IFN-β) and several varieties of human Le interferon (e.g., Hu IFN-α_1, Hu IFN-α_2). Up to 10 different varieties of Hu IFN-α have been identified. The nucleotide sequence has been determined for the Hu IFN-β gene and a number of the Hu IFN-α genes, and from this nucleotide sequence one could deduce the complete amino acid sequence (Table 2). Both Hu IFN-β and Hu IFN-α_1 appear to consist of 166 amino acids, whereas Hu IFN-α_2 comprises 165 amino acid residues. The amino acid sequence shown in Table 2 represents the mature polypeptide. *Statu nascenti*, this polypeptide would be preceded by a signal peptide (of 21 amino acids for Hu IFN-β, 23 amino acids for Hu IFN-α_1; and at least 17 amino acids for Hu IFN-α_2). During transport of the nascent protein across the endoplasmic membrane the signal peptide would be cleaved off, as is the rule for

Table 2 Amino Acid Sequence of Human Fibroblast Interferon (Hu IFN-β) and Human Leukocyte Interferon (Hu IFN-α_1 and Hu IFN-α_2)[a]

Hu IFN-β :						Met	Ser	Tyr	Asn	*Leu*	Leu	Gly	Phe	Leu	Gln	Arg	Ser	*Ser*	Asn	Phe
Hu IFN-α_1 :								Cys	Asp	*Leu*	Pro	Glu	Thr	His	Ser	Leu	Asp	Asn	Arg	Arg
Hu IFN-α_2 :								Cys	Asp	*Leu*	Pro	Gln	Thr	His	Ser	Leu	Gly	*Ser*	Arg	Arg
	Gln	Cys	Gln	Lys	*Leu*	Leu	Trp	Gln	Leu	Asn	Gly	Arg	*Leu*	Glu	Tyr	*Cys*	*Leu*	*Lys*	*Asp*	*Arg*
	Thr	Leu	Met	Leu	*Leu*	Ala	Gln	Met	Ser	Arg	Ile	Ser	Pro	Ser	Ser	*Cys*	*Leu*	Met	*Asp*	*Arg*
	Thr	Leu	Met	Leu	*Leu*	Ala	Gln	Met	Arg	Arg	Ile	Ser	*Leu*	Phe	Ser	*Cys*	*Leu*	*Lys*	*Asp*	*Arg*
	Met	Asn	*Phe*	Asp	Ile	*Pro*	Glu	*Glu*	Ile	Lys	Gln	Leu	Gln	*Gln*	*Phe*	*Gln*	*Lys*	Glu	Asp	*Ala*
	His	Asp	*Phe*	Gly	Phe	*Pro*	Gln	*Glu*	Glu	Phe	Asp	Gly	Asn	*Gln*	*Phe*	*Gln*	*Lys*	Ala	Pro	*Ala*
	His	Asp	*Phe*	Gly	Phe	*Pro*	Gln	*Glu*	Glu	Phe	–	Gly	Asn	*Gln*	*Phe*	*Gln*	*Lys*	Ala	Glu	Thr
	Ala	Leu	Thr	Ile	Tyr	*Glu*	*Met*	Leu	*Gln*	Asn	*Ile*	*Phe*	Ala	Ile	*Phe*	Arg	Gln	Asp	Ser	*Ser*
	Ile	Ser	Val	Leu	His	*Glu*	Leu	Ile	*Gln*	Gln	*Ile*	*Phe*	Asn	Leu	*Phe*	Thr	Thr	Lys	Asp	*Ser*
	Ile	Pro	Val	Leu	His	*Glu*	*Met*	Ile	*Gln*	Gln	*Ile*	*Phe*	Asn	Leu	*Phe*	Ser	Thr	Lys	Asp	*Ser*
	Ser	Thr	Gly	*Trp*	Asn	*Glu*	*Thr*	Ile	Val	Glu	Asn	Leu	Leu	Ala	Asn	Val	*Tyr*	His	*Gln*	Ile
	Ser	Ala	Ala	*Trp*	Asp	*Glu*	Asp	Leu	Leu	Asp	Lys	Phe	Cys	Thr	Glu	Leu	*Tyr*	Gln	*Gln*	Leu
	Ser	Ala	Ala	*Trp*	Asp	*Glu*	*Thr*	Leu	Leu	Asp	Lys	Phe	Tyr	Thr	Glu	Leu	*Tyr*	Gln	*Gln*	Leu
	Asn	His	*Leu*	Lys	Thr	Val	Leu	Glu	Glu	Lys	Leu	Glu	Lys	Glu	Asp	Phe	Thr	Arg	Gly	*Lys*
	Asn	Asp	*Leu*	Glu	Ala	Cys	Val	Met	Gln	Glu	Glu	Arg	Val	Gly	Glu	Thr	Pro	Leu	Met	Asn
	Asn	Asp	*Leu*	Glu	Ala	Cys	Val	Ile	Gln	Gly	Val	Gly	Val	Thr	Glu	Thr	Pro	Leu	Met	*Lys*
	Leu	Met	*Ser*	Ser	*Leu*	His	Leu	*Lys*	Arg	*Tyr*	Tyr	Gly	*Arg*	*Ile*	Leu	His	*Tyr*	*Leu*	*Lys*	Ala
	Ala	Asp	*Ser*	Ile	*Leu*	Ala	Val	*Lys*	Lys	*Tyr*	Phe	Arg	*Arg*	*Ile*	Thr	Leu	*Tyr*	*Leu*	Thr	Glu
	Glu	Asp	*Ser*	Ile	*Leu*	Ala	Val	Arg	Lys	*Tyr*	Phe	Gln	*Arg*	*Ile*	Thr	Leu	*Tyr*	*Leu*	Lys	Glu
	Lys	Glu	*Tyr*	*Ser*	His	*Cys*	*Ala*	*Trp*	Thr	Ile	*Val*	*Arg*	Val	*Glu*	*Ile*	Leu	*Arg*	Asn	*Phe*	Tyr
	Lys	Lys	*Tyr*	*Ser*	Pro	*Cys*	*Ala*	*Trp*	Glu	Val	*Val*	*Arg*	Ala	*Glu*	*Ile*	Met	*Arg*	Ser	Leu	Ser
	Lys	Lys	*Tyr*	*Ser*	Pro	*Cys*	*Ala*	*Trp*	Glu	Val	*Val*	*Arg*	Ala	*Glu*	*Ile*	Met	*Arg*	Ser	*Phe*	Ser
	Phe	Ile	Asn	Arg	*Leu*	Thr	Gly	Tyr	*Leu*	*Arg*	Asn									
	Leu	Ser	Thr	Asn	*Leu*	Gln	Glu	Arg	*Leu*	*Arg*	Arg	Lys	Glu							
	Leu	Ser	Thr	Asn	*Leu*	Gln	Glu	Ser	*Leu*	*Arg*	Ser	Lys	Glu							

[a]As deduced from the nucleotide sequence of the cloned interferon genes. The amino acids in italics are identical for Hu IFN-β and Hu IFN-α_1 (or -α_2).

proteins intended for secretion. There is considerable homology between human Le interferon and human F interferon: 49 of the 166 amino acid positions of F interferon and Le interferon (Hu IFN-α_1) are identical, and for Hu IFN-α_2 55 of the 165 amino acid positions are identical to those found in F interferon. The longest stretches of contiguous conserved amino acids are Gln-Phe-Gln-Lys (positions 47 to 50 of Hu IFN-α_1, 46 to 49 of Hu IFN-α_2, and 49 to 52 of Hu IFN-β) and Cys-Ala-Trp (positions 139 to 141, 138 to 140, and 141 to 143, respectively); for Hu IFN-α_2 and Hu IFN-β there is even a common stretch of five contiguous conserved amino acids: Cys-Leu-Lys-Asp-Arg (positions 29 to 33 and 31 to 35, respectively). It is likely that at least some of the conserved amino acids are essential for the activity of interferon.

The amino acid composition of human F interferon is strikingly similar to those of human Le interferon (Table 3), human Ly interferon, and mouse interferon. All these interferons are characterized by large quantities of leucine, glutamic acid-glutamine, and aspartic acid-asparagine, and, especially, by the abundance of hydrophobic amino acids (leucine, isoleucine, valine, phenylalanine). The latter account for almost one-third of all amino acid residues, which seems consistent with the notorious hydrophobicity of the interferon molecules. There are six proline residues in Hu IFN-α_1, five in Hu IFN-α_2, but only one in Hu IFN-β (Table 3). There are five cysteine residues in Hu IFN-α_1, four in Hu IFN-α_2, and three in Hu IFN-β, which seems consistent with the concept that S-S bonds are required for the biologic activity of interferon.

As mentioned above, human Le interferon and human Ly interferon are not identical. A comparison of the 35 amino terminal amino acids of human Ly interferon (determined by direct amino acid sequence analysis) and human Le interferons Hu IFN-α_1 and Hu IFN-α_2 (both obtained by cloning) has revealed nine differences between Ly interferon and Hu IFN-α_1 and five differences between Ly interferon and Hu IFN-α_2, which suggests that these three interferons are probably encoded by three nonallelic genes. Human Ly interferon has been designated Hu IFN-α_3.

Thus, interferon represents a rather heterogeneous class of molecules. Within a given class of interferons (i.e., human leukocyte interferon, mouse interferon) the differences in size may, at least in part, be attributed to differences in the extent of glycosylation. For example, the 35,000 dalton (or 38,000 dalton) component of mouse interferon can be identified as the more extensively glycosylated form of the 22,000 dalton species. In the presence of glycosylation inhibitors such as D-glucosamine or 2-deoxy-D-glucose, the cells produce primarily the 22,000 dalton species and little, if any, 38,000 dalton interferon, and in the presence of tunicamycin, a more drastic glycosylation

Table 3 Amino Acid Composition of Human Fibroblast Interferon (Hu IFN-β) and Human Leukocyte Interferon (Hu IFN-α_1 and Hu IFN-α_2)[a]

Amino acid	Hu IFN-β	Hu IFN-α_1	Hu IFN-α_2
Leu	24	22	21
Cys	3	5	4
Asn	12	6	4
Arg	11	12	10
Phe	9	8	10
Pro	1	6	5
Gln	11	10	12
Lys	11	8	10
Ala	6	10	8
Glu	13	15	14
Ile	11	7	8
Ser	9	13	14
Trp	3	2	2
Tyr	10	4	5
Val	5	6	7
Asp	5	11	8
Thr	7	9	10
Gly	6	3	5
Met	4	6	5
His	5	3	3
Total	166	166	165

[a]As deduced from the nucleotide sequence of the cloned interferon genes.

inhibitor, they produce an interferon species of even smaller size (18,000 daltons). The 38,000 dalton species can be converted to the 22,000 dalton species by oxidative cleavage with acid-periodate buffer. The 22,000 dalton interferon is apparently also embellished with some carbohydrate since it can be further cleaved to a 15,000 dalton species. However, the latter interferon species cannot be considered as a native interferon, since it is not produced as such by the cells. For human leukocyte interferon, the differences in molecular size between the two major species (16,000 and 21,000 daltons) would also reside in nonuniform glycosylation. No such variation has so far been noted for human fibroblast interferon. Like type I interferons, type II interferons may also be considered as glycoproteins, although it has not been clarified how much the carbohydrate moiety contributes to the heterogeneity of type II interferons.

Which is the role of the carbohydrate component in the biologic activity of interferon? (1) The carbohydrate residues are apparently dispensable for the antiviral and for many, if not all, of the other biologic effects of interferon, as has been demonstrated with mouse interferon and human Le interferon: interferon molecules that have been stripped of their carbohydrate residues retain their full antiviral activity. (2) The presence of carbohydrate residues is not required for the secretion of interferon by the cells. (3) Neither are they essential for the antigenicity of interferon, since nonglycosylated and fully glycosylated interferon (namely, human leukocyte interferon) are neutralized equally well by antisera to native interferon. (4) The carbohydrate residues may impart stability to the interferon molecules, since deglycosylated human leukocyte interferon is less thermostable than its glycosylated counterpart. (5) The carbohydrate residues may modulate the interaction of the interferon molecules with their cellular receptor(s), thereby confining the range of host cells receptive to interferon. In this sense, the carbohydrate component may be held responsible for what has generally been referred to as the species specificity of interferon. Nonglycosylated (or poorly glycosylated) interferons, (i.e., the 16,000 dalton species of human leukocyte interferon) are more cross-reactive (more active on heterologous cells) than are the fully glycosylated interferons (i.e., the 21,000 dalton species of human leukocyte interferon). Thus, the "core" or protein moiety would guarantee the antiviral (and other biologic) activities of interferon, whereas the carbohydrate moiety would restrict these activities to the cells of the same species from which the interferon was derived. (6) The carbohydrate residues may contribute to the well-known hydrophobicity of the interferon molecules, i.e., by displacing hydrophobic amino acid residues from the interior to the exterior of the molecule. (7) Finally, the carbohydrate residues seem to play an important role in the binding of interferon to a variety of ligands such as blue dextran, polyribonucleotides [i.e., poly(U) and poly(I)], concanavalin A, and other lectins (i.e., *Wistaria floribunda* agglutinin) which have all been applied successfully in the purification of interferon.

PURIFICATION OF INTERFERON

For many years progress in interferon research has been seriously hampered by the difficulties encountered in purifying interferon preparations. Even when produced under optimal conditions, interferon represents no more than 0.001 to 0.01% of the starting material, and attempts to separate it from the other molecules have often proven unsuccessful merely because interferon, when finding itself alone in solution, seeks something to cling or stick to. However, by the judicious choice and combination of various purification techniques, several interferons, including human fibroblast, leukocyte, and lymphoblastoid interferon, as well as mouse interferon, have recently been purified to apparent homogeneity. That the product was indeed homogeneous could be demonstrated by sodium dodecyl sulfate (SDS)-polyacrylamide gel electrophoresis which revealed the presence of a single protein band (human fibroblast interferon) or two protein bands (mouse interferon, human leukocyte interferon), coinciding with the expected antiviral (or anticellular) activity of the molecule. From the migration of these protein bands in the gels one could estimate the apparent molecular weights (Table 1), and with the fractions eluted from the gel one could determine the specific activity of interferon. For all pure type I interferons obtained so far the specific activity is in the vicinity of 10^9 antiviral interferon units per milligram of protein (~ 2 to 10×10^8 U/mg for the human interferons; $\sim 2 \times 10^9$ U/mg for mouse interferon).

The two main factors that have contributed to the total purification of interferon are the development of new purification techniques, primarily based on affinity chromatography, and the recognition that interferon had to be produced on a large scale if pure interferon were ever to be acquired. As reviewed in Table 4, several chromatographic procedures have been applied successfully to the purification of both type I and type II interferons, although purification to homogeneity has so far not been achieved for type II interferon. The binding of interferon to such ligands as blue (dextran) Sepharose, concanavalin A-Sepharose and poly (U)-Sepharose is governed by the so-called polynucleotide-binding site of the molecule. This polynucleotide-binding site would be present in both type I interferon and type II interferon and would, in turn, depend on the presence of the carbohydrate moiety. For example, the 21,000 dalton glycosylated species of human leukocyte

Table 4 Interferon Purification Procedures

Blue Sepharose CL-6B (Cibacron Blue F3GA-agarose)
Poly(U)- or poly(I)-Sepharose 4B
Concanavalin A-Sepharose
Phenyl-Sepharose CL-4B
Zinc chelate (iminodiacetate)-Sepharose 6B
Controlled pore glass beads
Affi-Gel 202
Conventional gels (Sephadex G-50 or G-100, Carboxymethyl-Sephadex, Bio-Gel P-60, P-100, P-150 or P-200, and so on)
Antibody (preferably monoclonal antibody) coupled to a solid matrix (Sepharose)
High-performance liquid partition chromatography (Lichrosorb RP-8)
SDS-PAGE (sodium dodecyl sulfate-polyacrylamide gel electrophoresis)

interferon is retained by blue dextran Sepharose, poly(U)-Sepharose, and concanavalin A-Sepharose. It is endowed with the polynucleotide-binding domain. On the contrary, the 16,000 dalton unglycosylated species of human leukocyte interferon is not retained by either blue dextran, poly(U), or concanavalin A, since it does not possess the polynucleotide-binding area. Similarly, human leukocyte interferon does not attach to the polynucleotide ligand if it has been produced in the presence of glycosylation inhibitors, such as tunicamycin, D-glucosamine, or 2-deoxy-D-glucose, and the 21,000-dalton species of human leukocyte interferon loses its affinity for concanavalin A after the carbohydrate moiety has been removed by treatment with periodate or glycosidases.

Some of the ligands listed in Table 4 have proven quite efficacious in the purification of interferon and through the sequential use of two or more of these ligands it has been possible to obtain pure interferon:

Mouse interferon by sequential affinity chromatography on poly(U)-Sepharose and anti-interferon globulin coupled to Sepharose; a more elaborate scheme involves precipitation with ammonium sulfate, chromatography on carboxymethyl-Sephadex, treatment with blue dextran and polyethylene glycol, gel filtration on Bio-Gel P-60 and Bio-Gel P-200, chromatography on phosphocellulose, isoelectric focusing, and, finally, chromatography on octyl-Sepharose.

Human fibroblast interferon by affinity chromatography on blue Sepharose followed by SDS-polyacrylamide gel electrophoresis; alternative purification systems involve controlled pore glass beads followed by zinc chelate (iminodiacetate)-Sepharose 6B chromatography; or concanavalin A-Sepharose followed by phenyl-Sepharose chromatography; or controlled pore glass beads followed by concanavalin A-Sepharose chromatography.

Human leukocyte interferon by a series of trichloroacetic acid precipitations, followed by Sephadex G-100 chromatography and high-performance liquid chromatography (with Lichrosorb RP-8).

Human lymphoblastoid interferon by chromatography on anti-interferon globulin (coupled to Sepharose), Sephadex G-150, SP-Sephadex C-25, and L-tryptophyl-L-tryptophan-Affi-Gel 10, and, finally, SDS-polyacrylamide gel electrophoresis.

All these interferons concern type I interferon. For the purification of type II interferons similar affinity sorbents have been applied, e.g., concanavalin A-Sepharose, Affi-Gel 10, Affi-Gel 202, blue Sepharose, poly(U)-Sepharose, and phenyl-Sepharose. One of the most successful purification schemes of type II interferon involves sequential chromatography on controlled pore glass and poly(U)-Sepharose. However, neither human nor murine type II interferon has so far been obtained in homogeneous form. The highest specific activity yet reported for type II interferon was 10^6 (antiviral) U/mg of protein. Since type II interferon is less effective as an antiviral agent than type I interferon, complete purification may be attained at specific activities lower than those reported for pure type I interferons (2×10^8 to 2×10^9 U/mg).

The more complicated the purification scheme, the lower the final recovery of interferon activity. Ideally, the whole purification process should not comprise more than two steps, or, if at all possible, only one step. Such complete purification in a single step may be achieved by using monoclonal antibody as the immunoadsorbent.

INTERFERON INDUCERS

Interferon can be induced by a wide variety of substances, including viruses, bacteria, bacterial components (i.e., endotoxins), some antibiotics (i.e., cycloheximide), double-stranded polyribonucleotides [i.e., $(I)_n \cdot (C)_n$], polycarboxylates (i.e., pyran copolymer), and synthetic low molecular weight compounds (i.e., tilorone). These substances are known to induce type I interferon. Type II interferon is produced by lymphocytes (T lymphocytes only ?) in response to either mitogens or specific antigens (Table 5).

Table 5 Interferon Inducers

Inducers of type I (classic) interferon
Microorganisms
Viruses (including animal, bacterial, fungal, mycoplasmatal, insect, and plant viruses)
Bacteria
Chlamydiae
Rickettsiae
Mycoplasms
Protozoa
Microbial components
Double-stranded RNAs (extracted from viruses or virus-infected cells)
Glycoproteins (from enveloped viruses such as Sendai or influenza virus)
Endotoxins (from bacteria)
Glutarimide antibiotics (from *Streptomyces*; e.g., cycloheximide, 9-methylstreptimidone)
Synthetic high molecular weight compounds (polyanions)
Polycarboxylates [e.g., pyran (maleic anhydride divinyl ether) copolymer, polyacrylic acid (PAA), and chlorite-oxidized oxyamylose (COAM)]
Polysulfates (e.g., polyvinylsulfate, λ carrageenan)
Polyphosphates (e.g., dextran phosphate)
Polynucleotides [e.g., $(I)_n \cdot (C)_n$: polyinosinic acid · polycytidylic acid]
Synthetic low molecular weight compounds
Fluorene and fluorenone derivatives (e.g., tilorone: 2,7-bis[2-(diethylamino)ethoxy]fluorenone)
Anthraquinone derivatives
Pyrazolo[3,4-b]quinoline derivatives
Acridine derivatives
Propanediamine derivatives
Pyrimidine derivatives (e.g., 2-amino-5-bromo-6-methyl-4-pyrimidinol and 2-amino-5-bromo-6-phenyl-4-pyrimidinol)
Inducers of type II (immune) interferon
T- (or B-) lymphocyte-stimulating agents
Mitogens (e.g., phytohemagglutinin, concanavalin A, staphylococcal enterotoxin A, and pokeweed mitogen)
Mixed lymphocyte cultures
Antilymphocyte antibody
Specific antigens [e.g., tuberculin or PPD (purified protein derivative), streptolysin O, tetanus toxoid, diphtheria toxoid, tumor cells, viral antigens]
Enzymes [e.g., galactose oxidase (through oxidation of galactose residues at the lymphocyte membrane)]

Source: With permission from S. Karger A. G. Basel: *Antibiotics Chemother.* *27*:251-287, 1980.

The most proficient inducers of interferon are viruses and double-stranded RNAs. They are able to induce interferon in various cell systems, both in vitro (cell culture) and in vivo (animals). Endotoxins are only effective in vivo and so are polycarboxylates and low molecular weight compounds. The interferon induction process of endotoxin, polycarboxylates, and low molecular weight compounds may involve both macrophages (as the inducer-reactive cells) and lymphocytes (as the interferon-producing cells). A similar two-step mechanism may be operative in the production of immune interferon, although type II interferon inducers (i.e., phytohemagglutinin) might also stimulate the production of intefferon in those lymphocytes that react to them. Type II interferons inducers are effective in vitro and in vivo.

Viruses

Almost all viruses, including single- and double-stranded RNA viruses and double-stranded DNA viruses, have proven capable of stimulating interferon production. In view of the excellent interferon-inducing activity of double-stranded RNAs, it is logical to attribute the interferon-inducing potentials of double-stranded RNA viruses (i.e., reovirus, blue tongue virus, rice dwarf virus, *Penicillium chrysogenum* mycophage) to their double-stranded RNA content. For single-stranded RNA viruses, whether + RNA (i.e., Semliki forest virus, sindbis virus) or - RNA (Newcastle disease virus, vesicular stomatitis virus), it would seem more difficult to allocate the actual interferon-inducing moiety. Some authors have attributed the interferon-inducing activity of single-stranded RNA viruses to their input RNA. It is more likely, however, that not the input RNA but a transcriptive (double-stranded RNA) intermediate may serve as the inducer of interferon. Indeed, ± RNA defective-interfering (D_I) particles, which contain covalently linked self-complementary + message and - anti-message RNA, appear to be extremely efficient inducers of interferon, and a single double-stranded RNA molecule per cell may be the minimum amount required to induce the synthesis of interferon. The molecule does not have to be double stranded over its whole length: half to one turn of a double helix, comprising 6 to 12 base pairs, may suffice to trigger the interferon response. Thus, a low threshold level of viral RNA synthesis, and not the entire transcriptive intermediate, would be required for the induction of interferon. Obviously, this limited viral RNA synthesis may be accomplished under conditions which do not lead to virus replication or progeny production. Carrying on this argument one step further, one might also speculate that DNA viruses (i.e., pox-, herpes-, and adenoviruses) owe their interferon-inducing properties to the formation of double-stranded RNA fragments. These double-stranded RNA fragments may be limited in size. Possibly, they may constitute no more than one helical turn, and correspond to those viral mRNA segments that are eliminated during the splicing process.

Bacteria and Bacterial Components

The interferon-inducing capacity of bacteria is primarily associated with their lipopolysaccharide (endotoxin) content. For some bacteria (i.e., *Klebsiella*), interferon induction would reside in the polysaccharide moiety. For other bacteria (i.e., *Salmonella*), however, it would reside in the lipid A portion. For yet other bacteria (i.e., *Brucella*), two active principles have been identified which should be recombined to confer full interferon-inducing activity. At present, it is rather difficult to speculate on a common chemical entity that would be responsible for the interferon-inducing activity of all bacteria. A characteristic difference between the patterns of interferon induction by bacteria and bacterial endotoxins on the one hand, and viruses, on the other, is that with bacteria and endotoxins maximum interferon production is obtained within 2 h (after intravenous administration of the inducer to mice or rabbits), whereas virus-induced interferon attains its peak titer only at 12 h after injection. A notable exception to this rule is *Brucella abortus*, which, much like viruses, elicits a late interferon response (in mice).

Polycarboxylates

The ability of polycarboxylates, i.e., pyran copolymer, polyacrylic acid, and chlorite-oxidized oxyamylose (COAM), to induce interferon resides in their polyanionic structure, characterized by a high molecular weight (exceeding 10,000), a high density of negative charges, and a stable backbone, which limits their degradation by the host. The anionic groups do not have to be carboxyl groups, since polyanions other than polycarboxylates, i.e., polysulfates and polyphosphates, are also notorious for their interferon-inducing ability. Interferon induction by polycarboxylates has almost exclusively been shown in mice, and the compounds should be delivered by the intraperitoneal route to yield the greatest efficacy; in this case serum interferon titers reach their peak values at about 18 to 24 h after drug administration. For polycarboxylates, interferon production could merely be regarded as the consequence of an overall stimulatory effect on macrophage activity.

Low Molecular Weight Compounds

Unlike other interferon inducers, low molecular weight compounds, such as tilorone (2,7-bis[2-(diethylamino)ethoxy]fluorenone) and ABMP (2-amino-5-bromo-6-methyl-4-pyrimidinol) are exquisitely effective upon oral administration. The only exception to this rule are the propanediamine derivatives. As with the polycarboxylates, interferon induction by low molecular weight inducers has almost exclusively been shown in mice, and peak serum interferon titers generally occur 18 to 24 h after drug administration. For ABMP, however, interferon peaks at 6 to 12 h, and this inducer is also effective in cats, as is ABPP (2-amino-5-bromo-6-phenyl-4-pyrimidinol). What is typical of the low molecular weight inducers of interferon is that, despite the lack of structural similarity among the different classes of these inducers (e.g., fluorenes, fluorenones, anthraquinones, pyraozolo[3,4-b]quinolines, acridines), there is, within each class a rather strict limitation of chemical modifications if the interferon-inducing potency is to be retained.

Double-Stranded Polyribonucleotides

By far the most potent inducers of interferon are the double-stranded RNAs, whether these are of biologic origin (i.e., reovirus, blue tongue virus) or synthetic origin [i.e., $(I)_n \cdot (C)_n$ or polyinosinic acid · polycytidylic acid (Fig. 2)]. They stimulate the formation of interferon in a wide variety of cell cultures (i.e., human diploid fibroblasts), animals (rabbits, mice, cats, dogs, calves, monkeys) and humans. In contrast with polycarboxylates and low molecular weight compounds, $(I)_n \cdot (C)_n$ induces a relatively early interferon response, peaking at 2 h (rabbits), 6 h (cats), or 12 h (humans). The salient requirements for its interferon-inducing ability are a high molecular weight (exceeding 30,000), high thermal stability ($T_m \geqslant 60°$), adequate resistance to enzymatic degradation, double strandedness as opposed to single or triple strandedness (although the molecule does not have to be double helical over its whole length), and the presence of free 2'-hydroxyl groups in both strands, or at least the pyrimidine strand. In attempts to increase the efficacy of $(I)_n \cdot (C)_n$ as an interferon inducer, various $(I)_n \cdot (C)_n$ analogs have been synthesized. The chemical modifications that appeared compatible with interferon-inducing activity are listed in Figure 2. By virtue of their increased resistance to nucleases, some of these $(I)_n \cdot (C)_n$ analogs, namely, $(I)_n \cdot (s^2C)_n$, $(I)_n \cdot (br^5C)_n$, and $(dIfl)_n \cdot (C)_n$, may even turn out to be more effective interferon inducers than the parent compound, particularly in those animal species that show a large capacity to hydrolyze $(I)_n \cdot (C)_n$. For the same reasons, one may expect higher interferon yields with a $(I)_n \cdot (C)_n$ derivative that has been complexed to poly-L-lysine and carboxymethylcellulose. There might be conditions, however, in which rapid degradation of $(I)_n \cdot (C)_n$ would be desirable to prevent the toxic side effects of the compound. In this case one may appeal to the mismatched analogs of $(I)_n \cdot (C)_n$, namely $(I_x,U)_n \cdot (C)_n$ and $(I)_n \cdot (C_xU)_n$ ($x \geqslant 10$). These mismatched $(I)_n \cdot (C)_n$ analogs are almost as effective in inducing interferon as

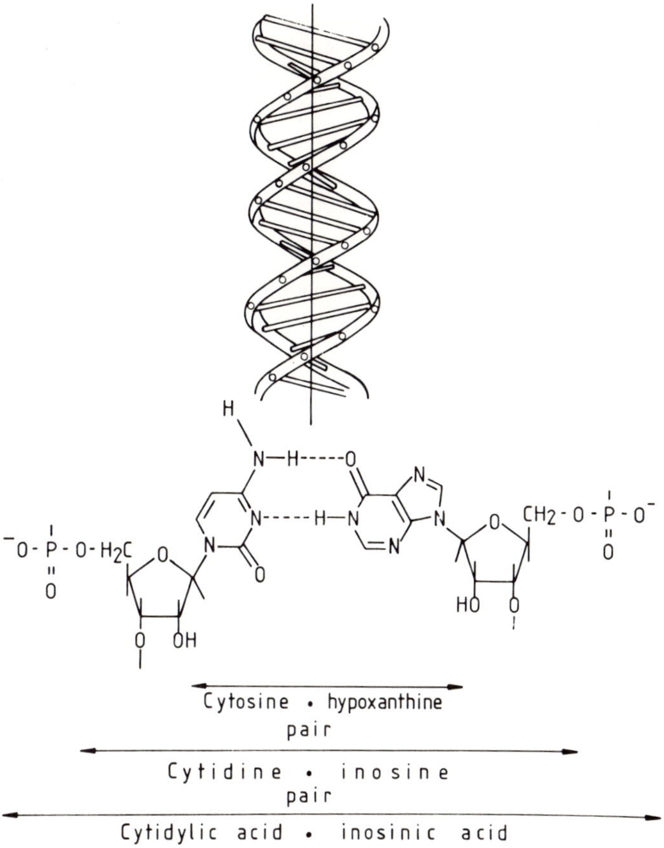

Figure 2 Polyinosinic acid • polycytidylic acid $[(I)_n \cdot (C)_n]$. Modifications which are compatible with interferon inducing activity: Substitution of sulfur for oxygen at C_2 of cytosine: $(I)_n \cdot (s^2C)_n$. Substitution of bromine for hydrogen at C_5 of cytosine: $(I)_n \cdot (br^5C)_n$. Substitution of azido, fluorine, or chlorine for hydroxy at C_2' of inosine: $(dIn_3) \cdot (C)_n$, $(dIfl)_n \cdot (C)_n$, $(dIcl)_n \cdot (C)_n$. Substitution of thiophosphate for phosphate: $(\bar{s}I)_n \cdot (C)_n$, $(I)_n \cdot (\bar{s}C)_n$, $(\bar{s}I)_n \cdot (\bar{s}C)_n$. Substitution of uracil for hypoxanthine, or uracil for cytosine: $(I_x,U)_n \cdot (C)_n$ and $(I)_n \cdot (C_x,U)_n$, provided $x \geqslant 10$. (With permission from S. Karger AG, Basel: *Antibiotics Chemother.* 27:251-287, 1980.)

$(I)_n \cdot (C)_n$ itself, but, as they are more susceptible to enzymatic degradation, they may not persist long enough in biologic fluids to elicit toxic side effects.

Type II ("Immune") Interferon Inducers

Type II interferon can be induced in sensitized lymphocytes by specific antigens [tuberculin or PPD (purified protein derivative)] and in unsensitized lymphocytes by mitogens [i.e., phytohemagglutinin (PHA), concanavalin A (ConA), and staphylococcal enterotoxin A (SEA)]. Since the mitogens (PHA, ConA, SEA) commonly used to induce type II interferon are T-cell mitogens, type II interferons have also been referred to as T-type interferons. The type II interferon induced by tuberculin may also be considered a T-type interferon, since tuberculin is assumed to be a T-dependent specific antigen. The question now arises whether the immune interferons induced by B-cell mitogens (i.e., pokeweed mitogen) or B-dependent specific antigens (i.e., endotoxin) still qualify as true type II interferons. These B-type interferons differ in some properties (i.e., pH 2 sensitivity, heat stability, antigenicity) from T-type interferons and, therefore, belong to a separate class of type II (B-type) interferons. According to some authors, B-type interferon might even be classified as type I interferon. Another point that has not been established yet is whether the T-type interferon induced by T-cell mitogens (i.e., PHA) is identical to the T-type interferon induced by T-cell-specific antigens (i.e., PPD). Although type II interferon inducers are operative in vivo and in vitro, their interferon-inducing properties have almost uniquely been pursued in vitro (lymphocyte cultures). In vitro interferon production attains its peak value after varying times (1 to 8 days), depending on a number of factors such as the immune status of the host and the nature of the mitogenic or antigenic stimulus. Although "immune" interferon may originate from either T or B lymphocytes, intimate contact with macrophages is required for maximum interferon production, and this interferon production is accompanied by the appearance of a whole series of mediators of cellular immunity ("lymphokines"), including MIF (migration inhibitory factor) and LT (lymphotoxin).

Considering the diversity of substances recognized as interferon inducers (Table 5), one may wonder how all these substances succeed in triggering such a highly specialized cellular function as interferon formation. Obviously, there is no structural relationship among the different classes of interferon inducers. Yet, within each class, little or no substitutions are allowed if the interferon-inducing ability is to be retained. Perhaps, a common denominator all interferon inducers share is that they are capable of interacting with a membranous component. This component may be located at the outer or inner cell membrane, and, when triggered, it may release a message that would lead to a derepression of the cellular genome and the expression of the interferon gene.

MECHANISM OF INTERFERON INDUCTION

The major events of the interferon induction process have been deciphered with either $(I)_n \cdot (C)_n$ or NDV (Newcastle disease virus) as the model inducer. The principles deduced from the use of these two inducers may probably apply to other interferon inducers as well. The actual trigger of the interferon induction process may well be the interaction of the inducer with a specific receptor site at the outer or inner cell membrane (Fig. 3). Since antibodies to double-stranded RNA $[(I)_n \cdot (C)_n]$ recognize many of the structural features responsible for the interferon-inducing capacity of double-stranded RNAs,

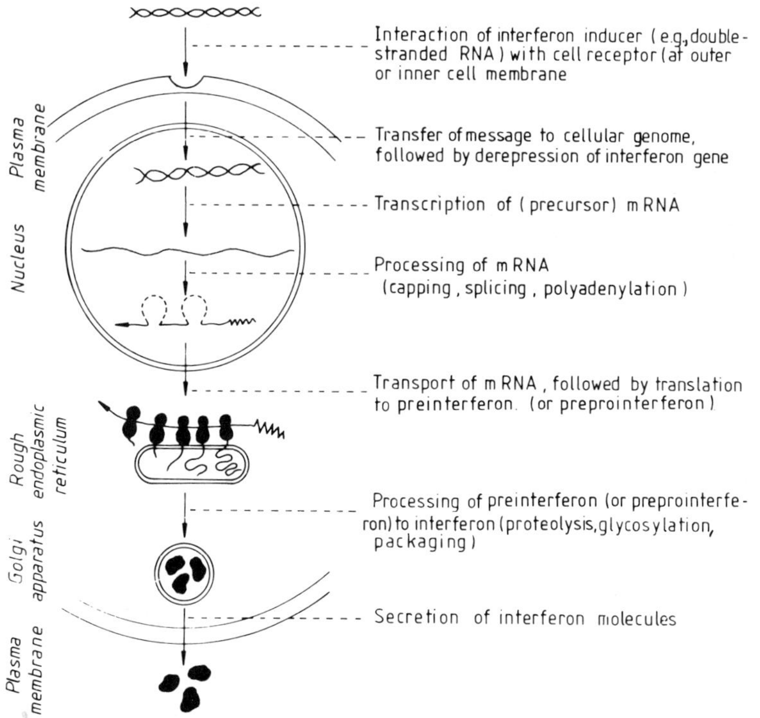

Figure 3 Mechanism of interferon induction.

one may postulate that the cellular receptor site for these double-stranded RNAs is a protein. For $(I)_n \cdot (C)_n$ the receptor site may be located at the outer cell membrane; for the double-stranded transcriptive intermediate of NDV it may be located inside the cell. After the receptor has been triggered by the interferon inducer, a signal is transmitted to the cellular genome, and as a result, the interferon gene is switched on (Fig. 3). However, direct evidence for this stage of the interferon induction process is lacking. The fact is that within a few hours after exposure of the cells to $(I)_n \cdot (C)_n$ or NDV, interferon mRNA accumulates in the cell. Interferon mRNA can be extracted from the induced cells, and, when incubated with heterologous cells or cell extracts (i.e., rabbit reticulocyte lysates, wheat germ extracts) or when injected into *Xenopus laevis* oocytes, it will direct the synthesis of biologically active interferon molecules. As has been demonstrated for various other eukaryotic mRNAs, interferon mRNA may be processed by capping, splicing, and polyadenylation (Fig. 3). The extent and sequence of these posttranscriptional modifications has not yet been established. Once it has been transported to the cytoplasm, interferon mRNA will associate with membrane-bound ribosomes (rough endoplasmic reticulum) and give rise to a primary translation product (preinterferon or preprointerferon?) which may be further modified by proteolytic cleavage to interferon. Before being released by the cell, the interferon molecules would be glycosylated, which, if analogous to other glycoproteins, should occur at the serine, threonine, or asparagine residues of the interferon molecule. The degree of glycosylation may vary considerably from one interferon species to another (i.e., F interferon versus Le interferon), and when interferon is produced in the presence of glycosylation inhibitors (i.e., D-glucosamine, 2-deoxy-D-glucose, tunicamycin), the carbohydrate moiety is either reduced in size or totally absent.

When human or rabbit fibroblast cultures are exposed to $(I)_n \cdot (C)_n$, interferon production becomes detectable by 1 h, rises to a peak by 3 h and is shut off by 6 to 8 h. However, when inhibitors of RNA synthesis (i.e., actinomycin D, 5,6-dichloro-1-β-D-ribofuranosylbenzimidazole) or protein synthesis (i.e., cycloheximide) are added to the cell cultures at the appropriate dosage and at the appropriate time (that is, after the interferon inducer), interferon production does not stop and the interferon yields may be increased up to 1000-fold. This phenomenon is generally referred to as interferon "superinduction," and has proven extremely useful in the large-scale production of human fibroblast interferon. It now appears that at least two mechanisms would contribute to the paradoxical enhancement of interferon production in the presence of metabolic inhibitors. First, interferon mRNA would be more stable in the presence of the inhibitors: during an ordinary induction interferon mRNA would decay with a half-life of approximately

0.5 h, but during superinduction its half-life may be extended to 6 to 8 h. Second, the apparent rate of synthesis of interferon mRNA could be increased in the presence of metabolic inhibitors, i.e., 3- to 4-fold, if the cell cultures are maintained in the presence of cycloheximide and 5,6-dichloro-1-β-D-ribofuranosylbenzimidazole. The first mechanism would imply the existence of a repressor protein that acts at the translation level; this repressor may be induced coordinately with interferon. It would normally inactivate interferon mRNA, thereby terminating interferon production, but in superinduced cells, no repressor would be synthesized and the interferon mRNA would remain functional for a much longer time, thus allowing the synthesis of interferon in increased yields. The second mechanism may involve a repressor protein that operates at the transcription level; this repressor may be a preexisting, rapidly turning over protein that normally inhibits transcription of the interferon gene. Inducers such as $(I)_n \cdot (C)_n$ would inactivate this repressor (either directly or indirectly), thus leading to the transcription of the interferon gene, and the presence of inhibitors of RNA or protein synthesis during induction may further lower the levels of this labile repressor, thus effecting an increased transcription rate relative to that observed in inhibitor-free induced cell cultures. Hence, two repressor substances may control the interferon induction process, a transcriptional repressor which prevents the expression of the interferon gene in the uninduced cell, and a translational repressor which shuts off interferon synthesis in the induced cell. Obviously, neither of these putative repressors has been isolated so far.

The interferon induction situation is further complicated by the fact that on induction of one particular cell strain (i.e., human diploid fibroblasts) by one particular inducer (i.e., NDV) both F (fibroblast) and Le (leukocyte) interferon may be produced. Similarly, human lymphoblastoid cells may produce some F interferon in addition to Ly (lymphoblast) interferon after they have been triggered with the appropriate inducer (i.e., Sendai virus). Human fibroblast cells induced with $(I)_n \cdot (C)_n$ synthesize only F interferon, but after stimulation with NDV, up to 20% of the total amount of interferon produced by the fibroblast cultures is Le interferon. Although human Le interferon and human F interferon represent different entities, both physically and antigenically, the kinetics of Le and F interferon production by human fibroblasts are remarkably similar. There is also a close correspondence in the shut-off of the synthesis of the two interferons, which suggests that, although Le and F interferon are encoded by two separate genes, their induction is controlled by a similar, if not identical, derepression mechanism.

While it was originally reported that chromosomes 2 and 5 were necessary together for human interferon production or may even carry separate human interferon genes, more recent studies indicate that neither chromosome 2 nor 5 seem to be involved, but that the gene for human interferon can be assigned to chromosome 9. This conclusion was reached upon an extensive analysis of various human-mouse cell hybrid clones. In these studies NDV served as the interferon inducer, and the human interferon produced was believed to be of the F type. If so, one may indeed contend that a gene on chromosome 9 is involved in the production of human fibroblast interferon. This does not necessarily mean that human leukocyte interferon is also encoded by a gene located on chromosome 9. Nor does it exclude the possibility that genes on chromosomes other than chromosome 9 are also required for human interferon production.

MASS PRODUCTION OF INTERFERON

For clinical use vast amounts of interferon are required, and in view of the species specificity or restricted host range of interferon, interferon should be produced by human cells, if it is to be used in humans. Various types of human cells, i.e., leukocytes, amnion cells, and diploid fibroblasts, as well as lymphoblastoid, fibroblastoid, and epithelial cell lines, could be monitored to produce large quantities of interferon, but the interferon preparations that have been regularly used in clinical investigations are the so-called leukocyte (Le) and fibroblast (F) interferons. Other equally important sources of human interferon are fibroblastoid, lymphoblastoid, and immune (or type II) interferon. Fibroblastoid interferon is probably identical to F interferon; lymphoblastoid (Ly) interferon is similar but not identical to Le interferon, while immune interferon may be considered as a totally different variant of interferon (Table 6).

The source of human *leukocyte interferon* are the leukocyte buffy coats derived from blood donated for transfusion. The leukocytes are separated from the other blood components by centrifugation, further purified, and then induced by Sendai virus (or Newcastle disease virus) to yield interferon. The leukocytes of one 400 ml blood bag can produce 1 to 2 million U of interferon, which is about the amount of interferon required to treat one patient for 1 day. The production of human leukocyte interferon cannot be scaled up readily, owing to the limited supply of human blood; and this logistic problem has turned the attention of investigators to alternative interferon sources.

One such source is human *lymphoblastoid interferon*, derived from lymphoblastoid cell lines which can be maintained in suspension culture for indefinite periods of time and which produce relatively large interferon quantities (up to 10 U of interferon per 10^3 cells)

Table 6 Human Interferon Preparations[a]

Interferon type	Cell source	Inducer	Availability	Safety
Type I				
Leukocyte (Le)	Peripheral blood leukocytes (from buffy coats)	Sendai virus (or another paramyxovirus)	Limited (depends on the supply of buffy coats)	May contain pyrogens and lymphokines other than interferon
Lymphoblastoid (Ly)	Lymphoblastoid cell lines (e.g., Namalwa)	Sendai virus (or another paramyxovirus)	Expansion easy, due to prolific growth of cells in suspension	Questionable (since Namalwa cells contain integrated Epstein-Barr virus sequences)
Fibroblast (F)	Diploid fibroblasts (i.e., derived from skin biopsy)	$(I)_n \cdot (C)_n$ (or another double-stranded RNA)	Expansion possible but expensive (since cells grow in monolayer)	May be pyrogenic (even after extensive purification)
Fibroblastoid (F)	Aneuploid fibroblasts (i.e., derived from tumor tissue)	$(I)_n \cdot (C)_n$ (or another double-stranded RNA)	Expansion possible but expensive (since cells grow in monolayer)	Questionable (since cells are tumor derived)
Type II				
Immune	Peripheral blood lymphocytes (from buffy coats)	Staphylococcal enterotoxin A (a T-cell mitogen)	Limited (depends on the supply of buffy coats)	May contain pyrogens and lymphokines other than interferon

[a]Alternative sources of human interferon are genetically engineered bacteria (*E. coli*), which have been found to express the activity of several interferon species, i.e., leukocyte interferon (Hu IFN-α_1 and -α_2) and fibroblast interferon (Hu IFN-β). These bacterially derived interferons may be made available in unlimited amounts. They should be carefully purified from bacterial contaminants before their clinical use could be envisaged.

when induced with Sendai virus, Newcastle disease virus, or measles virus. The most widely used of these lymphoblastoid cell lines is Namalwa, a cell line that has been established from a Burkitt's lymphoma patient. If suboptimal interferon yields are obtained they may be further elevated by pretreating the cells with short-chain fatty acids, i.e., propionic acid, n-butyric acid, or n-valeric acid. Although an undeniably expensive product, human lymphoblastoid interferon can be produced at a truly large scale, should its clinical use be called for.

Human *fibroblast interferon* is produced by diploid fibroblast cell strains derived from human skin, foreskin, fetal skin and muscle tissue, or embryonic lung. Fibroblast cell strains have a limited life span, allowing no more than 50 to 70 cell doublings. In optimal conditions, that is, when "primed" with small amounts of interferon (100 U/ml), induced by $(I)_n \cdot (C)_n$ [or an appropriate $(I)_n \cdot (C)_n$ analog], and subsequently superinduced with metabolic inhibitors, such as cycloheximide and actinomycin D, human fibroblasts may readily produce 30 interferon U per 10^3 cells. Although the production of F interferon can be expanded if necessary, mass propagation of fibroblast cells is hampered by the fact that these cells, unlike lymphoblastoid cells, do not grow in suspension but only when attached to a suitable surface. The production of F interferon is generally performed with fibroblast cultures grown in glass or plastic roller bottles, but it has been suggested that interferon yields (per milliliter culture medium) can be increased if the cells are cultured on microcarrier dextran beads instead of roller-bottle walls.

Aneuploid fibroblastoid cell lines derived from malignant tumors constitute another useful source of F interferon, the more that some of these cell lines, i.e., MG-63 (derived from osteogenic sarcoma), are better producers of interferon than normal fibroblasts [when induced with $(I)_n \cdot (C)_n$]. Unlike fibroblast cells, fibroblastoid cell lines can be maintained in cell culture for undefinite periods of time. The interferon produced by these established cell lines is usually referred to as *fibroblastoid interferon*. However, current evidence suggests that it may be identical to fibroblast (F) interferon. Like lymphoblastoid interferon, fibroblastoid interferon may, because of its tumor cell origin, fail to satisfy safety requirements for clinical testing, unless it is purified to homogeneity.

Human *immune interferon* can be produced at a large scale by using peripheral blood lymphocytes as the cell substrate and staphylococcal enterotoxin A as the inducer. The yields are as high as those obtained for human leukocyte interferon, and although human interferon preparations would be devoid of leukocyte (type I) interferon activity, they may certainly contain a number of lymphokines other than interferon which should first be separated from immune interferon before the properties of the latter can be assessed.

Even if the conventional techniques for the mass production of human leukocyte, lymphoblastoid, fibroblast, fibroblastoid, and immune interferon were to be further refined, it is unlikely that the supply of interferon could keep up with clinical demands. To break the interferon production bottleneck, one should appeal to DNA recombination technology, insert the interferon gene in bacterial plasmids, and incite the bacteria to produce interferon. This has so far been accomplished with human fibroblast interferon (Hu IFN-β) and human leukocyte interferon (Hu IFN-α_1 and Hu IFN-α_2) and may eventually be achieved for the other interferon types, i.e., lymphoblastoid (Ly) interferon (Hu IFN-α_3) and type II (immune) interferon. One may expect that the interferon quantities delivered by bacterial systems will by far exceed those produced by conventional means. The DNA recombination approach toward interferon production comprises several steps: (1) the extraction and partial purification of interferon mRNA from induced cells; (2) the transcription of this mRNA to complementary single-stranded DNA (ssDNA) or further on to double-stranded cDNA; (3) the insertion of this cDNA into bacterial DNA, i.e., with plasmid pBR322 as vector and *Escherichia coli* as the bacterial host; (4) identification of those bacterial clones that contain the (full-length) interferon gene; and (5) as these clones may not spontaneously produce (sufficient quantities of) biologically active interferon, they should be further manipulated to increase their yields of interferon. While the DNA recombination approach permits a full characterization (including base sequencing) of the cloned interferon genes, the clinical usefulness of the bacterially produced interferon will ultimately depend on the role of the carbohydrate groups in the biologic activity of the interferon molecules. Indeed, bacteria do not afford glycosylation of the proteins they synthesize, and if carbohydrate groups do turn out to be essential, i.e., for the stability of the interferon molecules in biologic fluids, genetic engineering may not be the most profitable route toward interferon production, unless additional measures are taken to cure the lack of appropriate glycosylation.

ACTION OF INTERFERON AT THE MOLECULAR LEVEL

Interferon is not active in the cell in which it has been induced. It must first move out of the cell and diffuse to neighboring cells, where it will interact with a specific receptor at the cell surface. This interaction somehow triggers the synthesis of a number of cellular proteins. At least three proteins (apparent molecular weights 120,000, 80,000, and 67,000) are induced in mouse cells, and in human fibroblasts, at least five proteins (apparent molecular weights 120,000, 88,000, 80,000,

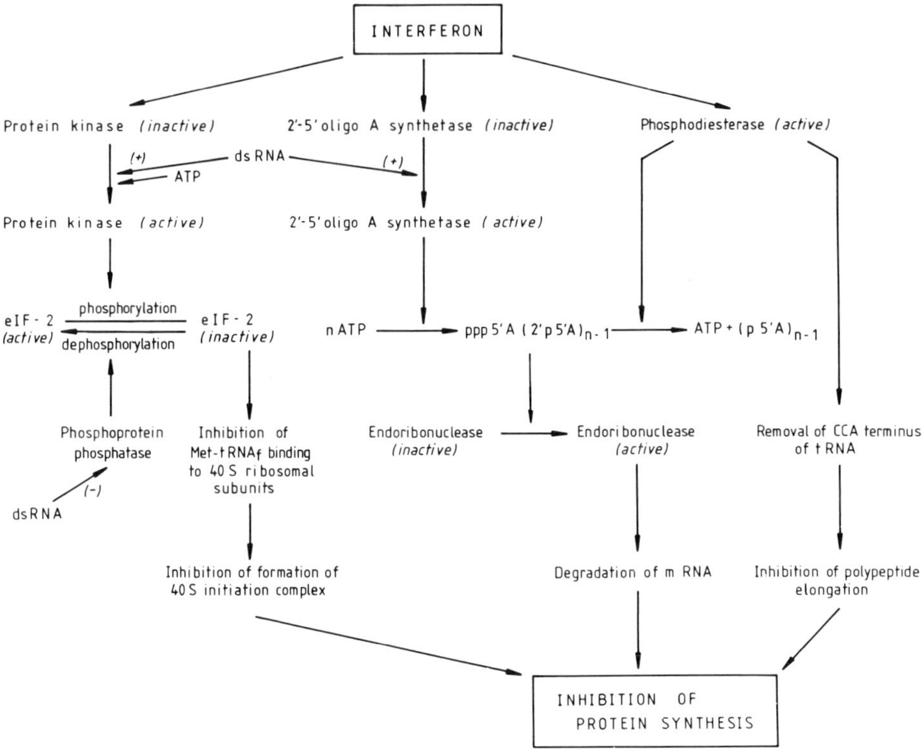

Figure 4 Cascade model of interferon action at the molecular level.

67,000, and 56,000) are induced. The induction of these proteins is blocked by actinomycin D if added together with interferon but not if added 2 h after interferon. Likewise, the establishment of the antiviral state becomes resistant to inhibition by actinomycin D within 2 h after the addition of interferon. In this respect, the appearance of the interferon-induced proteins correlates well with the development of the antiviral response to interferon. The induction of these proteins is transient: they reach a peak value within a few hours, which is followed by a decline in spite of the continuous exposure of the cell to interferon. Thus, the cell is committed only for a few hours to the expression of the interferon-induced proteins. The reason for the decay (e.g., shutoff of mRNA transcription or translation, inactivation of intermediary products) has not been elucidated.

For two of the interferon-induced proteins, the biologic role has been established. These are a protein (phospho)kinase and a 2'-

5'oligo A synthetase. Both are induced as inactive enzymes but are converted to their active form in the presence of double-stranded RNA (Fig. 4). This requirement may help to restrict the antiviral action of interferon to the virus-infected cell, since the latter provides the replicative dsRNA intermediate necessary for activation of the protein kinase and 2'-5'oligo A synthetase. Once it is activated, the protein kinase will transfer the γ-phosphate group from ATP to a suitable acceptor protein, i.e., the 67,000 (murine)-73,000 (human) dalton ribosomal protein and the 35,000 to 38,000 dalton subunit of eIF-2 (eukaryotic initiation factor 2). This phosphorylation leads to an inactivation of eIF-2. The process is reversible, however. The phosphorylation of eIF-2 is regulated by a phosphoprotein phosphatase, which is constitutively present in normal as well as interferon-treated cells. The phosphoprotein phosphatase is inhibited by dsRNA. In the absence of dsRNA it will cause a dephosphorylation of eIF-2 and hence convert eIF-2 to its active form. The initiation factor eIF-2 is required for the binding of methionyl-initiator tRNA (Met-$tRNA_f$) to 40S ribosomal subunits, and thereby initiates protein synthesis (Fig. 4). The role of eIF-2 in the biologic activity of interferon is further attested by the fact that the interferon-induced block of translation can be reversed readily by addition of eIF-2.

A second pathway by which interferon may inhibit protein synthesis (Fig. 4) involves the induction of an oligo-isoadenylate (2'-5'oligo A) synthetase, also designated as 2-5A synthetase (molecular weight, 56,000). Once it has been activated by dsRNA, 2-5A synthetase catalyzes the formation from ATP of isoadenylate oligomers with 2'-5' phosphodiester instead of the usual 3'-5' phosphodiester linkages. The isoadenylate oligomers formed by 2-5A synthetase comprise ppp5'A2'p5'A, ppp5'A2'p5'A2'p5'A (Fig. 5), and so on; they can be designated by the general formula $ppp5'A(2'p5'A)_n$, whereby n = 1 to 12. The isoadenylate oligomers then activate a preexisting endoribonuclease (RNAse) that breaks down RNAs, including mRNA (Fig. 4). Activation of this RNAse is reversible and lost upon removal of 2'-5'oligo A. However, the enzyme can be activated again by reexposure to 2'-5'oligo A. Its molecular weight is 185,000, unusually large for a nuclease. Activation of the RNAse is not accompanied by a substantial change in either molecular size or conformation.

Thus, both the protein kinase and 2'-5'oligo A synthetase pathways ultimately result in the inhibition of protein synthesis (Fig. 4). Both pathways are applicable equally well to interferon type I (i.e., human leukocyte interferon, human fibroblast interferon) as interferon type II. Interferon may inhibit protein synthesis through yet a third enzymatic pathway, that is, a 2'(3')-5' phosphodiesterase. The activity of this phosphodiesterase is not controlled by dsRNA. The enzyme seems to play a dual role. (1) It removes the -CpCpA terminus of tRNA, thus

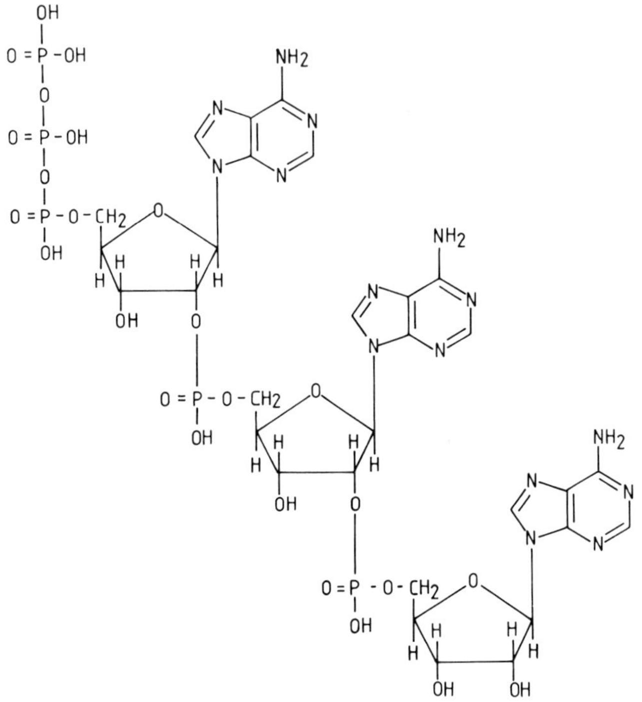

Figure 5 ppp5'A2'p5'A2'p5'A.

preventing the aminoacylation of tRNA. The reduced translation efficiency that results from this inhibitory effect can be partially overcome by the addition of tRNA. (2) The phosphodiesterase antagonizes the 2'-5'oligo A synthetase pathway by degrading 2'-5'oligo A (Fig. 4). This would explain why the activation of the RNAse and the concomitant degradation of mRNA is only a transient phenomenon.

The 2'-5'oligo A-activated ribonuclease shows no mRNA specificity and degrades cellular and viral mRNAs to the same extent. Thus, it is not immediately clear how the 2'-5'oligo A synthetase pathway could lead to a preferential inhibition of viral protein synthesis. Nor is it obvious how the protein kinase and phosphodiesterase pathways could discriminate between viral and cellular protein synthesis. However, since both the protein kinase and 2'-5'oligo A synthetase pathway depend on the presence of dsRNA, one may postulate that these pathways are only operational in close proximity to the replicative dsRNA intermediate formed during virus replication. The fact that 2-5A synthetase is efficiently incorporated into the virion is also suggestive of some association between the enzyme and specific viral functions. In the virus-infected cell, 2-5A synthetase may not be evenly distributed but rather be accumulated in those compartments in which virus assembly takes place. As a result of the subcellular location of dsRNA and 2-5A synthetase, the synthesis of 2'-5'oligo A and the activation of the ribonuclease may only occur at the sites of virus replication, which may explain why the RNAse could preferentially degrade viral mRNA, leaving cellular mRNA unaffected.

If introduced artificially in permeabilized cells, 2'-5'oligo A would be capable of inhibiting protein, RNA, and DNA synthesis; these effects are transient in uninfected cells after treatment with low concentrations (< 20 nM) of 2'-5'oligo A. In the virus-infected cell, however, the shutdown of cellular metabolism is not reversed upon removal of 2'-5'oligo A; the virus-infected cell dies with or without treatment with 2'-5'oligo A. By 2'-5'oligo A treatment, however, viral replication is depressed or arrested. A similar sequence of events may occur spontaneously in the interferon-treated virus-infected cells, as the amounts of 2'-5'oligo A produced in these cells are sufficient to cause an inhibition of protein synthesis. However, the induction of 2'-5'oligo A is not always associated with an antiviral or anticellular effect. For example, undifferentiated embryonal carcinoma (EC) cells respond to interferon by an increased 2'-5'oligo A synthetase activity, although these cells are refractory to both the antiviral and anticellular effects of interferon. Even more remarkable is the behavior of HEC-1, a clonal transformed human cell line. These cells constitutively express protein kinase and 2'-5'oligo A synthetase activity, yet they do not respond to the antiviral and anticellular activity of interferon. One may infer, therefore, that the induction of 2'-5'oligo A synthetase and protein kinase is probably a necessary but certainly not a sufficient condition for the establishment of an antiviral or anticellular state.

In most instances the inhibitory effects of interferon on virus replication could ensue from a primary or secondary block of viral protein synthesis, resulting in the shutoff of "early" viral proteins (i.e., RNA or DNA polymerases) and "late" viral proteins (i.e., capsid proteins), respectively. For some viruses (namely, retroviruses), however, the interferon-induced block may be situated at a very late stage of virus morphogenesis, presumably at the level of virus assembly or release. The production of retrovirus (i.e., murine leukemia virus) proteins is not inhibited by interferon, and virus particles are being assembled, but they are not released from the cell. Instead, they seem to accumulate at the cell surface, and after they have eventually been released they are not capable of infecting other cells. The reduced infectivity of these virus particles might be due to a defect in the proteolytic cleavage normally associated with virion assembly. Suggestive of such defective assembly is that retrovirus particles produced by interferon-treated cells contain, in addition to the usual structural proteins, an aberrant glycoprotein of a rather large molecular weight (85,000 daltons).

Vesicular stomatitis virus particles produced by interferon-treated cells have also reduced infectivity. This low infectivity may be related to the reduced amounts of glycoprotein (G) and membrane protein (M) incorporated into such particles. It is possible that changes in the cell plasma membrane may be the basis for the alteration in infectivity of both vesicular stomatitis virus and murine leukemia virus produced by interferon-treated cells because the morphogenesis of these viruses involves budding from the cell surface. There are, however, alternative explanations. For example, the defectiveness of the virions may be due to the incorporation into the virion of an interferon-induced cellular protein, i.e., 2-5A synthetase, which interferes with the normal replicative cycle of the virus. Inside the virion, 2-5A synthetase is probably inactive, but following uncoating of the virions within the cell, the synthetase may become active and initiate the sequence of events (Fig. 4) that leads to destruction of the viral mRNAs.

ACTION OF INTERFERON AT THE CELLULAR LEVEL

Since interferon exerts its antiviral effect by acting on the cell rather than on extracellular virus, there is no reason it should not exert other effects on cells, and indeed, interferon has been shown to induce a variety of biologic activities other than resistance to virus infection (Fig. 6). Quite schematically, these effects could be classified in three categories: (1) enhancement of certain specialized cell functions, (2) alteration of the cell surface, and (3) inhibition of cell division.

The following instances are representative of an enhancement of specialized cell functions: priming of the cells by interferon, so that the subsequent interferon response to a suitable interferon inducer is enhanced; increased production of some enzymes (i.e., aryl hydrocarbon hydroxylase after induction with benzanthracene); increased production of prostaglandin; increased release of histamine from basophils of allergic individuals after exposure to the allergen or anti-IgE antibody; increased cytotoxicity of sensitized T lymphocytes for target (tumor) cells; increased expression of histocompatibility (i.e., HLA and β_2-microglobulin) antigens on lymphoid cells; increased phagocytic, virucidal, and tumoricidal activity of macrophages; and activation of natural killer (NK) lymphocytes. The latter effect has received particular attention, since NK cells are assumed to play an important role in immune surveillance against spontaneous tumor development. NK cells, also referred to as FcR^+ cells since they express receptors for the Fc portion of IgG, are cytotoxic to tumor cells. It now appears that their

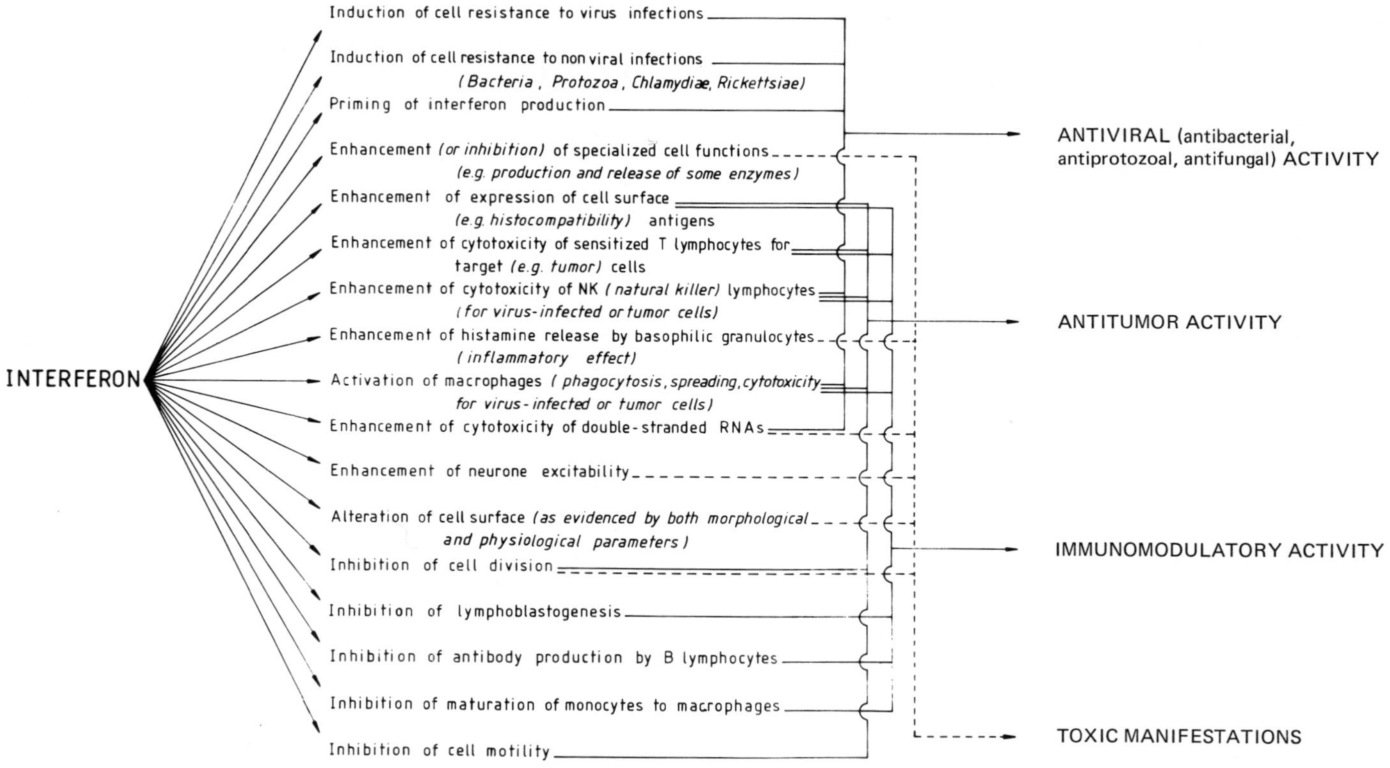

Figure 6 Variety of biologic effects of interferon (at cellular level).

cytotoxic activity can be enhanced by interferon treatment. This activation is time limited and reversible: the cytotoxicity of NK cells reverts to baseline values after the contact with interferon has ceased. However, the cells can be reactivated upon repeated exposure to interferon.

The increased expression of surface histocompatibility antigens in interferon-treated cells is an indication of a cell surface change generated by interferon. Other indications for such cell surface alterations include increased binding of (radiolabeled) concanavalin A, increased net negative charge of the cells (as determined by cell electrophoresis), decreased cell motility, diminished thymidine or uridine transport, reduced binding of cholera toxin and thyrotropin to plasma membranes, increased number of intramembranous granules, and increased density of the plasma membrane. Two other interferon effects, namely, increased excitability of nerve cells and increased susceptibility to double-stranded RNA toxicity, are also suggestive of cell surface modifications, the more so that these effects require only a short incubation time (30 min to 2 h) of the cells with interferon.

One of the most notorious effects that have been associated with interferon is an inhibition of cell division. Interferon is not cytotoxic. Its effect on cell proliferation should rather be regarded as a cytostatic event. This "anticellular" effect has been observed with both normal and transformed cells. At the level of the immune system the anticellular effect of interferon is reflected by an inhibition of B-cell proliferation, of antibody synthesis, and of lymphoblast transformation. Both the anticellular and antiviral activities of interferon would reside in the same molecule. The mechanism by which these activities are achieved may be dissimilar, however.

Many but not necessarily all of the effects generated by interferon may be secondary to the interaction of interferon with a specific cellular receptor. The exact nature of this interferon receptor has not yet been elucidated. The gene coding for this receptor has been tentatively assigned to chromosome 21 (in human cells), although direct evidence for this allocation is still lacking. One may wonder how a single molecule like interferon, after it has interacted with a (putatively) single receptor, could generate so many different biologic activities. However, these effects may not be as disparate as they seem at first glance. Most of the cellular alterations brought about by interferon could in fact be interpreted as manifestations of senescence. This cell senescence would be characterized by a thickening of the cytoskeleton, a decline in proliferative ability, a number of cell surface modifications (as the average cell size increases), and increased expression of histocompatibility antigens and some other cell functions.

Obviously, the protective activity achieved by interferon against viral infections in vivo will not only be determined by its direct antiviral effect but also by several other factors, such as a stimulating effect on the cytotoxicity of macrophages and NK lymphocytes for virus-infected cells (Fig. 6). It is even possible that, in vivo, such host-mediated effects may be more important in determining host resistance to virus infection than the direct inhibitory effect of interferon on virus replication. This may explain why interferon can be effective in vivo against some viruses which are not sensitive to its antiviral action in vitro. Likewise, the antitumor activity of interferon may depend on both a direct inhibition of tumor cell proliferation and an increased tumoricidal activity of T and NK lymphocytes and macrophages (Fig. 6). In its interaction with the immune system, interferon would display a modulatory activity, thereby stimulating some immune responses while obviating others: in general, the immunoenhancing effect of interferon would be achieved by an effect on T lymphocytes, in contrast with the immunosuppressive effect which would be mediated through an effect on B lymphocytes. As a result, interferon inhibits the primary and secondary antibody response to sheep red blood cells (in mice); it inhibits DTH (delayed-type hypersensitivity) and delays allograft rejection. It also inhibits the maturation of monocytes to macrophages. In as far as interferon stimulates the release of histamine, increases neuron excitability, inhibits cell division, makes cells more susceptible to the toxic action of double-stranded RNAs, and triggers yet other undesirable host cell responses, it may not only contribute to the recovery but also to many of the untoward manifestations of a virus disease (i.e., fever, malaise, fatigue, skin rash).

In addition to acquiring resistance to virus infection, interferon-treated cells are also capable of transferring this resistance to other neighboring cells by an as yet ill-defined cell-to-cell communication system. Thus, the action of interferon does not necessarily require a direct effect of interferon on each responding cell. Sensitive, fast-responding cells may transfer antiviral resistance to less sensitive, slower responding cells, thereby amplifying the activity of interferon. This transfer process has been demonstrated in vitro with both homogeneous and heterogeneous cell populations. It only concerns the transfer of antiviral resistance. Whether the immunoregulatory and other activities of interferon can be disseminated in a similar fashion remains to be established.

All the effects attributed to interferon (Fig. 6) can also be obtained with interferon inducers. It should be borne in mind, however, that in addition to interferon, inducers may elicit other biologic responses, which could either potentiate or counteract the effects obtained with interferon alone. For example, interferon inducers (such as double-stranded RNAs) may exert a direct inhibitory effect on cellular DNA, RNA, or protein synthesis, thereby complementing the cytostatic action of interferon itself. They may also exert a macrophage-activating effect on their own. Thus, interferon inducers may confer a greater antitumor activity than could be expected from the amounts of interferon they induce. On the other hand, interferon inducers generally act as immunoadjuvants, whereas interferon often suppresses the immune response; interferon inducers rather act as anti-inflammatory agents, whereas interferon promotes the inflammatory process; and finally, interferon inducers are potent mitogens for lymphocytes, whereas interferon itself curtails mitogenic activity. In view of the antagonistic nature of the latter activities, it may not always be possible to predict the efficacy of an interferon inducer as an antiviral or antitumor agent. Its efficacy may be dependent on a number of factors (i.e., immunoresponsiveness of the host) which are often beyond control.

EFFICACY OF INTERFERON IN EXPERIMENTAL SITUATIONS

As summarized in Table 7, interferon and interferon inducers have been found effective in inhibiting a number of experimental virus infections, whether the compounds were given locally, as in herpetic keratoconjunctivitis, or systemically, as in herpetic encephalitis. Inducers may be expected to have the same antiviral spectrum and potency as interferon itself, in so far as their effects are due solely to interferon production. There is little doubt that this is actually so, if the inducer is administered shortly (i.e., 24 h) before virus infection at doses which induce interferon levels equivalent to those attained by exogenous interferon administration. In this case, virus infection takes place at the height of the interferon response, and this may determine the final outcome of the disease. If, however, the inducer is administered some time (i.e., several days) before or after virus infection, its activity may deviate from that of interferon, since other factors, i.e., a direct stimulatory effect of the inducer on lymphocyte or macrophage activity, may now come into play. Interferon inducers may well give rise to higher peak interferon values than exogenously administered interferon, but due to hyporeactivity, these high values cannot be maintained, since a second dose of the inducer will invariably fail to stimulate the production of interferon if administered too shortly after the first dose. On the other hand, certain interferon inducers, i.e., polyacrylic acid and pyran copolymer, may remain protective long after they have ceased to stimulate the production of detectable interferon titers.

The most important drawback that both interferon and its inducers suffer from is that they must be given prophylactically to be fully effective. This means before or shortly (within 24 h or, at the most,

Table 7 Representative Experimental Virus Infections[a] in Which Interferon or Interferon Inducers [i.e., $(I)_n \cdot (C)_n$] Have Shown Efficacy

Virus	Disease	Host
Herpes simplex	Keratoconjunctivitis	Rabbit
	Encephalitis	Mouse
Cytomegalovirus	Systemic infection	Mouse
Vaccinia	Keratoconjunctivitis	Rabbit
	Tail lesions	Mouse
	Encephalitis	Mouse
	Skin lesions	Rhesus monkey
Coxsackie B	Myocarditis	Mouse
Foot-and-mouth disease	Encephalitis	Mouse
Encephalomyocarditis	Encephalitis	Mouse
Semliki Forest	Encephalitis	Mouse
Simian hemorrhagic fever	Encephalitis	Rhesus monkey
West Nile	Encephalitis	Mouse
Yellow fever	Hepatitis	Rhesus monkey
Influenza A, B	Respiratory infection	Mouse, baboon
Parainfluenza 1, 3	Respiratory infection	Mouse, hamster
Rabies	Encephalomyelitis	Mouse, rabbit, rhesus monkey
Vesicular stomatitis	Encephalitis	Mouse
Hepatitis B	Chronic hepatitis B	Chimpanzee

[a]Excluding virus-induced tumors.

48 h) after virus infection. Only in a very few instances interferon was found to exert a protective effect when given at 4 or 5 days after virus infection, at a time when virus had already multiplied in the target organ (brain) and symptoms (i.e., paralysis) begun to appear (i.e., in mice infected with vesicular stomatitis virus). However, to obtain a therapeutic effect with interferon, extraordinarily high doses (6.4×10^6 U per mouse) are required. Such doses do not seem feasible in humans. That interferon is merely a prophylactic antiviral agent

can easily be inferred from its biologic role during the course of a natural virus infection. Interferon can be considered as part of a first-line defense of the host against the viral intruder. In contrast with classic antibody, interferon is operational immediately after virus infection. It is produced at the site of infection, and will prevent further spread of the virus, either directly, or indirectly through the aid of newly recruited cells (i.e., NK lymphocytes or macrophages). An extraneous supply of interferon, as materialized by the administration of either interferon or interferon inducers, may help the host in its attempts to overcome the virus infection, but only if interferon is administered at the right time, that is, before the host has started to make its own interferon, and if the amounts of interferon made by the host are inadequate. One can expect little, if any, benefit from an interferon therapy that does not lead to higher interferon titers (either locally or systemically) than those which have been achieved by the virus itself. Nor is there much cause for administering interferon or interferon inducers at a time when both virus replication and endogenous interferon production are declining. Unfortunately, most virus diseases become manifest only when virus replication has passed its peak. In these conditions, interferon therapy would be of no avail.

The antiviral effects of interferon in vivo may be based upon a direct inhibition of virus replication, but in addition, they may also require the participation of different host cells, such as macrophages and NK lymphocytes. When activated by interferon, these host cells could selectively destroy virus-infected cells. That the efficacy of interferon as an antiviral agent needs the cooperation of the host is attested by several findings: (1) interferon can be effective in vivo against a virus insensitive to its antiviral action in vitro; and (2) the antiviral efficacy of interferon is drastically diminished in the immunosuppressed host: i.e., interferon is unable to protect congenitally athymic mice or ATS (anti-thymocyte serum)-treated mice against a lethal herpes simplex virus infection under conditions in which it confers significant protection in immunocompetent mice. These findings implicate the necessity of a fully competent T-cell immunity in the antiviral activity of interferon in vivo.

As an antiviral substance, interferon may be expected to inhibit the multiplication of oncogenic viruses as well as nononcogenic viruses, and, indeed, interferon has been found to inhibit the growth of various virus-induced tumors, including Rous sarcoma and Rauscher leukemia (Table 8). But in addition, interferon has also been found to inhibit the growth of a wide variety of transplantable murine ascitic or solid tumors of different origins (spontaneous, viral, or chemically induced) in different strains of mice. Furthermore, interferon has proven effective in inhibiting the growth of chemically and x-ray-induced tumors and in delaying the development of spontaneous tumors, such

Table 8 Experimental Tumor Models in Which Interferon or Interferon Inducers [i.e., $(I)_n \cdot (C)_n$] Have Shown Efficacy

Origin	Tumor	Host
Virus induced	Polyoma	Rat
	Simian virus 40	Hamster
	Human adenovirus 12	Hamster
	Shope fibroma	Rabbit
	Moloney sarcoma	Mouse
	Harvey sarcoma	Mouse
	Friend leukemia	Mouse
	Rauscher leukemia	Mouse
	Rous sarcoma	Chicken
	Radiation leukemia	Mouse
Chemically induced	9,10-Dimethylbenzanthracene induced	Mouse
	3-Methylcholanthrene induced	Mouse
X-ray induced	Radiogenic lymphoma	Mouse
Transplantable	L-1210 leukemia	Mouse
	Ehrlich ascites tumor	Mouse
	RC-19 ascites tumor	Mouse
	EL-4 ascites tumor	Mouse
	B-16 malignant melanoma	Mouse
	Mammary adenocarcinoma	Rat
	Lewis lung (3 LL) tumor	Mouse
	L-929 sarcoma	Mouse
	Walker 256 carcinosarcoma	Mouse
	B-1237 lymphoma	Mouse
	E ♂ G 2 tumor	Mouse
Spontaneous	AKR lymphoid leukemia	Mouse
	Mammary adenocarcinoma	Mouse

as lymphoid leukemia in AKR mice (Table 8). To delay the appearance of spontaneously developing tumors, interferon has to be given prophylactically, preferably from birth onward. Such prophylactic treatment is ineffective against transplanted tumors. Here, optimal antitumor effects are obtained if there is intimate contact between interferon and tumor cells, i.e., if both interferon and tumor cells are inoculated intraperitoneally. However, potent interferon preparations can also exert an inhibitory effect when inoculated at a site distant from that of tumor cell implantation, and in the case of Lewis lung carcinoma, interferon has been shown to inhibit the development of pulmonary metastases.

Various factors may contribute to the inhibitory effects of interferon on tumor growth, i.e., direct inhibition of tumor virus replication, direct inhibition of tumor cell multiplication, enhancement of expression of tumor cell surface antigens, enhancement of cytotoxicity of sensitized T lymphocytes for tumor cells, enhancement of cytotoxicity of macrophages for tumor cells, and enhancement of cytotoxicity of NK cells for tumor cells. Thus, the antitumor activity of interferon may be achieved by a direct action on the tumor cells, on the host's surveillor cells, or on both. When interferon was assayed for its antitumor effects in mice inoculated with either interferon-resistant L-1210 or interferon-sensitive L-1210 cells, both the interferon-resistant and interferon-sensitive L-1210 cells were inhibited by interferon treatment, although the growth of the interferon-sensitive cells was inhibited to a greater extent than the growth of the interferon-resistant cells. The fact that interferon inhibited the growth of the interferon-resistant cells suggests that this inhibition was mediated by the host; however, the fact that interferon preferentially inhibited the growth of the interferon-sensitive cells suggests that, in addition to the host-mediated effect, interferon acted directly on the tumor cells. To further distinguish between a direct inhibitory and host-mediated effect of interferon, one should evaluate the antitumor activity of both human interferon and mouse interferon in the human tumor-nude mouse system. Considering the defined host range ("species specificity") of interferons, which allows little, if any, cross-species activity between human and mouse interferon, one may reason that if the antitumor activity of interferon were due to a direct inhibitory effect on the tumor cell, human interferon but not mouse interferon would inhibit the growth of human tumors transplanted into nude mice. If, however, the antitumor activity of interferon would be based primarily on a host-mediated response (i.e., NK cell or macrophage activity), one may expect mouse interferon but not human interferon to be effective in inhibiting human tumor growth in nude mice. These assumptions are only valid in so far as (1) human interferon is incapable of activating the murine surveillor cells and (2) mouse interferon is not directly inhibitory to human tumor cells.

The antitumor activity of interferon is most pronounced if the tumor burden is low. Although interferon diminishes the growth of the tumor, a cure is not effected (unless fairly small numbers of tumor cells are injected), and a complete regression of a well-established tumor mass has never been observed under interferon treatment. This does not preclude that in the treatment of patients with malignancy, as in the treatment of patients with virus disease, interferon may prove to be of greater value than the results of animal experimentation suggest. Most of the experimental systems used to assess the antiviral and antitumor potentials of interferon are such that the virus and tumor grow rapidly and invariably lead to death of the animal within a minimum period of time. In patients the odds are less overwhelming in favor of the virus or tumor, so that in these conditions better results with interferon therapy may be looked for.

EFFICACY OF INTERFERON IN CLINICAL SITUATIONS

Several clinical studies indicate that interferon may have a beneficial effect in the therapy of some virus infections (Table 9). For example, human leukocyte and fibroblast interferon, when combined with minimal wiping debridement or trifluorothymidine, have been shown to accelerate the healing and to reduce the rate of recurrences in patients with dendritic keratitis. However, interferon alone does not seem to be sufficiently effective in topical treatment of this disease; it needs a strong therapeutic partner such as debridement or trifluorothymidine.

Table 9 Clinical Situations in Which Interferon Treatment Has Shown Efficacy[a]

Virus infections
Herpetic keratitis
Epidemic adenovirus keratoconjunctivitis
Herpes zoster or varicella in cancer patients
Cytomegalovirus infection in renal transplant recipients
Herpes labialis after surgery for "tics douleureux"
Chronic active hepatitis B (only in women?)
Acute respiratory tract (rhino)virus infections
Benign tumors
Laryngeal papilloma
Condyloma acuminatum
Malign tumors
Osteogenic sarcoma
Multiple myeloma
Non-Hodgkin's lymphoma
Advanced breast cancer
Acute lymphocytic leukemia
Hodgkin's disease
Nasopharyngeal carcinoma
Metastatic nodules of malignant melanoma

[a]For malignant tumors only in a few cases (only one case for Hodgkin's disease and nasopharyngeal carcinoma).

Favorable results have been reported with human fibroblast interferon in the local treatment of epidemic adenovirus keratoconjunctivitis: in the interferon-treated group the length of the disease was dramatically reduced, as was the number of cases developing keratitis.

In a randomized double-blind trial involving 90 cancer patients with early localized herpes zoster, human leukocyte interferon was found to prevent progression of the disease; interferon-treated patients had fewer days of new vesicle formation in the primary dermatome, a trend toward less acute pain, diminished severity of postherpetic neuralgia, and fewer visceral complications, as compared to the placebo-treated controls. In a similar trial conducted in children with varicella, human leukocyte interferon had no effect on the number of days of new vesicle formation, although there were fewer visceral complications in patients receiving interferon than in those receiving placebo. Human leukocyte interferon has been shown to reduce the incidence of cytomegalovirus viremia and to delay the excretion of cytomegalovirus in renal transplant recipients; however, there was no concomitant clinical improvement, and in parallel studies, human fibroblast interferon was found without effect on the overall incidence of virus infections in renal transplant recipients. In a limited number of patients who had undergone surgical decompression of the trigeminal sensory root for "tics douleureux" (trigeminal neuralgia), human leukocyte interferon prevented the appearance of herpes labialis if administered from the day before surgery onward.

While preliminary clinical trials suggested that human leukocyte and fibroblast interferon might be effective in the treatment of chronic active hepatitis B, as revealed by a diminution of several markers of the disease [e.i., Dane particle-associated DNA polymerase, hepatitis B core antigen (HBcAg), hepatitis B surface antigen (HBsAg)], this promising prospect was not borne out by a double-blind study with human leukocyte interferon. This study revealed no effect of interferon on the indices of hepatitis B virus infection, apart from a transient drop in DNA polymerase activity, which could be due to the fever that followed each interferon injection. Other studies pointed out that only female patients with chronic hepatitis B infection may respond to interferon treatment; male patients would not. The reason(s) for this sex difference remain(s) to be established.

Human leukocyte interferon administered by intranasal spray proved efficacious in the prevention of clinical symptoms and virus shedding in volunteers infected with rhinovirus 4. However, similar favorable effects have not yet been reported for other respiratory tract virus infections. In fact, intranasally applied human leukocyte interferon turned out to be inactive in the prophylaxis of an influenza B virus infection.

From the clinical data that have so far been reported it is difficult to assess the future of interferon as an antiviral drug. For those virus infections that have the greatest socioeconomic impact, i.e., respiratory tract virus infections, the indications that interferon would be effective are scarce at best. For other important virus infections, like chronic active hepatitis B, claims of efficacy have been disputed. Virus infections in transplant recipients do not appear an ideal indication either, since the antiviral potency of interferon may be seriously impaired in patients with a deficient immune system. The virus diseases which could benefit most from interferon therapy are the herpesvirus infections (i.e., herpetic keratitis, herpes zoster, herpes labialis), but here we have several new compounds at hand (i.e., acycloguanosine, bromovinyldeoxyuridine, fluoroiodoaracytosine, and phosphonoformate) that may supersede interferon in efficacy and that are after all easier and cheaper to manufacture.

The high hopes for interferon are not as much based on its potentials as an antiviral drug as on its potentials as an anticancer drug. Indeed, several reports, most of which are anecdotal, suggest that interferon may in some instances elicit a partial or even complete tumor regression (Table 9). The most comprehensive study on the role of human (leukocyte) interferon in the treatment of cancer has been conducted with osteosarcoma patients. An evaluation of the results after 2.5 years indicated that in the concurrent control group of patients 70% had developed pulmonary metastases and 65% had died, whereas in the interferon-treated group, 36% of patients had developed pulmonary metastases and 28% had died. After 5 years, 7 of 12 interferon-treated patients were still alive, whereas only 6 of 23 control patients were alive.

Multiple myeloma represents another malignant disease that appears to respond favorably to interferon therapy. Of a series of four myeloma patients given human leukocyte interferon, all four responded, two with a complete and two with a partial remission. In another study, only 3 of 10 myeloma patients responded to interferon therapy. Of a series of 11 non-Hodgkin's lymphoma patients treated with human leukocyte interferon, only 3 failed to respond; complete remission was noted in 2 patients with nodular poorly differentiated lymphocytic lymphoma. A similar tumor-reducing effect was achieved by human leukocyte interferon in another three patients with nodular poorly differentiated lymphocytic lymphoma. In contrast, no reduction in tumor size was observed in patients with rapidly advancing diffuse histiocytic lymphoma. Nor has an objective tumor regression been observed in non-small cell lung cancer patients treated with human leukocyte interferon. Of a series of 17 patients with advanced breast cancer who entered a trial with human leukocyte interferon, 6 demonstrated a partial remission,

1 improved slightly, and the other 10 failed to respond. Of five patients with acute lymphocytic leukemia, all five responded to human leukocyte interferon therapy with a drop in peripheral and marrow leukemic blast cells, and a complete remission has also been noted in a patient with acute lymphocytic leukemia following treatment with $(I)_n \cdot (C)_n$ (stabilized by poly-L-lysine and carboxymethylcellulose). When given to a patient with Hodgkin's disease, human leukocyte interferon caused a prompt tumor regression, but the remission turned out to be short-lived, since tumor growth resumed after a few months in spite of continuous interferon therapy.

Human fibroblast interferon has only occasionally been evaluated for its antitumor potentials. When injected directly into the tumor mass (cutaneous or subcutaneous metastatic lesions of two patients with malignant melanoma), human fibroblast interferon effected a marked reduction in tumor volume. Quite spectacular is the case of a boy with nasopharyngeal carcinoma, who had had two courses of high-dose radiotherapy and chemotherapy and relapsed while still on chemotherapy, the tumor having invaded his maxillary sinus, orbit, and brain: a complete regression of the tumor occurred after systemic treatment with human fibroblast interferon.

From these case reports one may deduce that interferon somehow interferes with malignant tumor growth, and as has been substantiated with laryngeal papilloma and condyloma acuminatum (Table 9), it may also reduce the development of benign tumors. However, many questions remain to be resolved before interferon could be accredited as an anticancer drug. These questions relate to the optimal dosage and duration of interferon treatment, the kinds of neoplastic disease amenable to interferon therapy, and the individual host factors that may influence the antitumor activity of interferon. Most of all, the efficacy of interferon should be compared with that of the conventional chemotherapeutic and radiotherapeutic means before its role in the treatment of cancer could be fully appreciated.

PROBLEMS INVOLVED IN INTERFERON THERAPY

In clinical trials interferon is routinely administered by the intramuscular route at a dose of 3 to 10×10^6 (antiviral) U/day. Higher dosage regimens are not feasible because of the toxicity of the presently available interferon preparations. This toxicity includes fever, myelosuppression, fatigue, and several other untoward effects, as reviewed in Table 10. In a representative study the incidence of the side effects of human leukocyte interferon was as follows: myelosuppression, 90%; fever, 80%; weight loss, 50%; pain at injection site, 30%; partial alopecia, 30%; rigors, 20%; headache, 10%; and skin rash, 10%. Hepatotoxicity

Table 10 Toxic Side Effects Noted upon Parenteral Administration of Human Leukocyte, Fibroblast, or Lymphoblastoid Interferon[a]

Myelosuppression (anemia, leukopenia, thrombocytopenia)
Fever, chills
Fatigue, malaise
Anorexia, weight loss
Nausea, vomiting
Rigors
Partial alopecia
Pain at injection site
Muscle weakness, diffuse myalgia
Confusion, depression
Skin rash (erythema)
Headache
Hypertension or hypotension
Liver dysfunction (as reflected by increased aminotransferase levels)

[a]Most side effects are equally applicable to all three types of human interferon.

and changes in blood pressure (hypo- or hypertension) were noted in some studies but not in others. All toxic effects were reversible and disappeared spontaneously when interferon treatment was stopped. The most common toxic manifestations were pyrexia and bone marrow suppression; in some cases they necessitated withdrawal of interferon. It is noteworthy that the symptoms associated with interferon therapy, i.e., fever, headache, anorexia, nausea, vomiting, tiredness, muscle weakness, confusion, also occur during the course of an acute virus infection. This implies that the symptoms of such infections may be caused by endogenous interferon production. Similarly, the adverse reactions observed after the administration of interferon inducers [i.e., $(I)_n \cdot (C)_n$, whether or not complexed with poly-L-lysine and carboxymethylcellulose] may to a large extent be mediated by the interferon induced, and should not necessarily be interpreted as toxicity innate to the inducer per se.

The toxic effects of interferon could be due to interferon itself or to impurities present in the interferon preparation. Indeed, the interferon preparations that have so far been used in clinical studies are very impure; only 0.1 to 1% of the protein they contain is actually interferon. Theoretically, the impurities may originate from the cell culture medium or serum (i.e., fetal calf serum proteins), from the inducer (i.e., Sendai virus proteins), from the cells (i.e., lymphokines, such as macrophage migration inhibitory factor, which are induced concomitantly with interferon), from the reagents used to purify interferon (i.e., concanavalin A) or to increase its yield (i.e., actinomycin D), or they may simply represent accidental contaminants. It is obvious that the factors coinduced with interferon, i.e., the skin reactivity factor coinduced with human fibroblast interferon, could explain some of the toxic effects associated with interferon therapy. For example, partially purified human fibroblast interferon was found to elicit both an immediate and a delayed type of skin reaction when injected intracutaneously. However, no such skin reactivity was seen with a completely pure interferon preparation, although the latter was still acting as a pyrogen when injected intravenously. It has become commonplace to attribute the deleterious effects of interferon preparations to the impurities and to allot their beneficial effects to interferon itself. Neither allegation seems correct. Interferon may have toxic properties of its own, and its clinical efficacy in the treatment of cancer and viral diseases might at least partially be accounted for by the impurities. Until completely pure interferon becomes available in amounts sufficient to conduct clinical trials, it will not be possible to resolve the question as to whether the various effects that have been recorded with interferon in humans are due to interferon or unidentified contaminants.

When used topically, as in the treatment of herpetic keratitis, less frequent administration of high-titered interferon is superior to more sustained treatment with lower doses. For systemic administration of interferon, the intramuscular or subcutaneous route is preferable. After intravenous injection, interferon is rapidly cleared from the bloodstream, and after oral administration, it does not even reach the bloodstream. Following intramuscular or subcutaneous administration, low but long-lasting interferon blood levels are obtained: approximately 200 U/ml over a period of 24 h after injection of 3×10^6 U of human leukocyte or lymphoblastoid interferon. Of the two major species of native human leukocyte interferon, the larger (21,000 dalton) species would be less efficient at entering the bloodstream and more rapidly cleared from the bloodstream than the smaller (15,000 or 16,000 dalton) species. Hence, the unglycosylated form of human leukocyte interferon would remain in circulation longer than the glycosylated form. And human fibroblast interferon, another fully glycosylated interferon, would not even achieve significant levels of activity in the bloodstream. One may infer, therefore, that the presence of carbohydrate residues in the interferon molecule somehow prohibits its appearance in the bloodstream. The mechanism(s) by which the carbohydrate alters the pharmacokinetic behavior of interferon has not yet been elucidated. Glycosylated interferons may fail to produce the expected levels of activity in the blood for some of the following reasons: (1) they may be more firmly bound at the injection site, or (2) more rapidly transported from the blood to the tissues, or (3) more rapidly inactivated in the blood. The last possibility can be eliminated, however, since human fibroblast interferon showed a similar clearance rate as human leukocyte interferon when both interferons were administered intravenously (to rabbits). Failure of human fibroblast interferon and other glycosylated forms of interferon to achieve adequate blood levels does not necessarily mean that they should be inferior to unglycosylated interferons in terms of therapeutic usefulness. It has been ascertained, for example, that human fibroblast interferon, in spite of its nonappearance in the bloodstream, is equally efficient as human leukocyte interferon in stimulating the activity of circulating NK cells.

While interferon could be held responsible for many of the subjective symptoms caused by a virus infection, it may also contribute to the objective manifestations of the disease, i.e., by propagating the inflammatory reaction. Interferon can do so by several mechanisms: (1) by enhancing the release of histamine from mast cells, and (2) by stimulating the production of prostaglandin E, itself a mediator of the inflammatory response. Inflammatory reactions play an important part in allergic and/or autoimmune diseases, such as rheumatoid arthritis, bronchial asthma, systemic lupus erythematosus, and ulcerative colitis. In these circumstances interferon therapy may be detrimental rather than beneficial. If interferon is administered to NZB (New Zealand black) mice, a mouse strain that spontaneously develops an autoimmune disorder resembling systemic lupus erythematosus, it accelerates progression of the disease. Interferon would also be responsible for the various manifestations of LCM (lymphocytic choriomeningitis) virus disease in mice infected at birth with LCM virus, including the appearance of glomerulonephritis later in life; the development of these manifestations of LCM virus disease can be inhibited by anti-mouse interferon globulin administered at the time of virus infection. Therefore, one may envisage some concrete situations, i.e., autoimmune processes, in which the use of anti-interferon globulin would be more advantageous than the use of interferon. Clearly, the exact conditions for the use of anti-interferon globulin need to be defined, but in view of the recent advances made in hybridoma technology (namely, production of monoclonal antibody of human origin), the prospects of an anti-interferon therapy may look perfectly feasible.

CONCLUSION

For more than 20 years interferon has seemed to hover on the edge of a breakthrough. Now at last there are hints that its hour may have come. What has suddenly caught public attention is the possible anticancer action of interferon. Indeed, interferon may have a tumor-reducing effect in at least some human cancers. This is not surprising, since interferon has also been reported to inhibit tumor cell growth in a wide variety of in vitro and in vivo systems.

For those who have been directly engaged in interferon research, interferon has been more often a source of frustration than of enchantment. Various factors have fostered these feelings of disappointment, not least the difficulties encountered in purifying interferon which made some people doubt the existence of the molecule, the multitude and disparity of the biologic effects that have been imputed to interferon, the limitations of the conventional cell culture procedures to produce interferon in quantities sufficient to keep pace with the clinical demands, and finally, the all but overwhelming results that have been obtained with interferon in some clinical studies. However, in the last one or two years many of these grim prospects have been remedied. Several interferon species, i.e., human leukocyte, fibroblast, and lymphoblastoid interferon, have now been purified to homogeneity. For human leukocyte and fibroblast interferon, the entire amino acid sequence has been deduced, which in turn opens new avenues for the chemical synthesis of the molecule or active portions thereof. The genes of human leukocyte interferon and human fibroblast interferon have been inserted into *E. coli* by genetic recombinant techniques, and since both genes were expressed in the bacterial cell, we can look forward to the large-scale manufacturing of human interferon by bacterial systems. Studies on the molecular mode of action have reached a high level of sophistication. While for many years the existence of a secondary antiviral protein (AVP) induced by interferon was considered as highly speculative, if not ludicrous, it now appears that interferon can induce a whole series of proteins that could all qualify as secondary antiviral proteins. Two of these proteins, the protein phosphokinase and 2'-5'oligo A synthetase, may have an important role to play in the mechanism by which interferon controls virus replication and cell growth.

Thus, the primary structure of interferon has to a large extent been unraveled, its mass production by DNA recombinant techniques is forthcoming, and refined hypotheses have been proposed to explain its mode of action in molecular terms. What has not become much clearer, however, is the therapeutic usefulness of interferon. Here, the question remains as to whether interferon is ever going to be used routinely in clinical medicine, and, if so, for what disease?

SUGGESTED READINGS

General Reviews

Baglioni, C. Interferon induced enzymatic activities and their role in the antiviral state. *Cell 17*:255-264, 1979.

Cantell, K. Why is interferon not in clinical use today? In *Interferon 1979*, Vol. 1 (I. Gresser, Ed.). Academic, London, 1979, pp. 1-28.

Carter, W. A. Bypassing the "species barrier" with carbohydrate altered interferon from leukocytes. *Cancer Res. 39*:3796-3798, 1979.

De Clercq, E. Effects of interferon on human tumor cell growth in nude mice. In *The Nude Mouse in Experimental and Clinical Research*, Vol. II (J. Fogh and B. Giovanella, Eds.). Academic, New York, 1982, pp. 439-449.

De Clercq, E. Interferon inducers. *Antibiot. Chemother. 27*:251-287, 1980.

De Maeyer, E. Interferon twenty years later. *Bull. Inst. Pasteur 76*:303-323, 1978.

Epstein, L. G. The comparative biology of immune and classical interferons. In *Biology of the Lymphokines* (S. Cohen, E. Pick, and J. J. Oppenheim, Eds.). Academic, New York, 1979, pp. 443-514.

Friedman, R. M. Antiviral activity of interferons. *Bacteriol. Rev. 41*:543-567, 1977.

Gresser, I. On the varied biologic effects of interferon. *Cell. Immunol. 34*:406-415, 1977.

Gresser, I., and M. G. Tovey. Antitumor effects of interferon. *Biochim. Biophys. Acta 516*:231-247, 1978.

Pollard, R. B., and T. C. Merigan. Experience with clinical applications of interferon and interferon inducers. *Pharmacol. Ther. A 2*:783-811, 1978.

Revel, M. Molecular mechanisms involved in the antiviral effects of interferon. In *Interferon 1979*, Vol. 1 (I. Gresser, Ed.). Academic, London, 1979, pp. 101-163.

Torrence, P. F., and E. De Clercq. Inducers and induction of interferons. *Pharmacol. Ther. A 2*:1-88, 1977.

Structure of Interferon

Derynck, R., J. Content, E. De Clercq, G. Volckaert, J. Tavernier, R. Devos, and W. Fiers. Isolation and structure of a human fibroblast interferon gene. *Nature 285*:542-547, 1980.

Mantei, N., M. Schwarzstein, M. Streuli, S. Panem, S. Nagata, and C. Weissmann. The nucleotide sequence of a cloned human leukocyte interferon cDNA. *Gene 10*:1-10, 1980.

., T., S. Ohno, Y. Fujii-Kuriyama, and M. Muramatsu. , nucleotide sequence of human fibroblast interferon cDNA. *Gene 10*:11-15, 1980.

Taniguchi, T., N. Mantei, M. Schwarzstein, S. Nagata, M. Muramatsu, and C. Weissmann. Human leukocyte and fibroblast interferons are structurally related. *Nature 285*:547-549, 1980.

Thang, M. N., D. C. Thang, M. K. Chelbi-Alix, B. Robert-Galliot, M. J. Commoy-Chevalier, and C. Chany. Human leukocyte interferon: Relationship between molecular structure and species specificity. *Proc. Natl. Acad. Sci. USA 76*:3717-3721, 1979.

Purification of Interferon

De Maeyer-Guignard, J., M. G. Tovey, I. Gresser, and E. De Maeyer. Purification of mouse interferon by sequential affinity chromatography on poly(U)- and antibody-agarose columns. *Nature 271*:622-625, 1978.

Grob, P. M., and K. C. Chadha. Separation of human leukocyte interferon components by concanavalin A-agarose affinity chromatography and their characterization. *Biochemistry 18*:5782-5786, 1979.

Knight, Jr., E., M. W. Hunkapiller, B. D. Korant, R. W. F. Hardy, and L. E. Hood. Human fibroblast interferon: Amino acid analysis and amino terminal amino acid sequence. *Science 207*:525-526, 1979.

Rubinstein, M., S. Rubinstein, P. C. Familletti, R. S. Miller, A. A. Waldman, and S. Pestka. Human leukocyte interferon: Production, purification to homogeneity, and initial characterization. *Proc. Natl. Acad. Sci. USA 76*:640-644, 1979.

Secher, D. S., and D. C. Burke. A monoclonal antibody for large-scale purification of human leukocyte interferon. *Nature 285*: 446-450, 1980.

Taira, H., R. J. Broeze, B. M. Jayaram, P. Lengyel, M. W. Hunkapiller, and L. E. Hood. Mouse interferons: Amino terminal amino acid sequences of various species. *Science 207*:528-529.

Zoon, K. C., M. E. Smith, P. J. Bridgen, C. B. Anfinsen, M. W. Hunkapiller, and L. E. Hood. Amino terminal sequence of the major component of human lymphoblastoid interferon. *Science 207*:527-528, 1980.

Interferon Inducers

De Clercq, E., G-F. Huang, B. Bhooshan, G. Ledley, and P. F. Torrence. Interferon induction by mismatched analogues of polyinosinic acid · polycytidylic acid $[(I_x,U)_n \cdot (C)_n]$. *Nucleic Acids Res. 7*:2003-2014, 1979.

De Clercq, E., B. D. Stollar, J. Hobbs, T. Fukui, N. Kakiuchi, and M. Ikehara. Interferon induction by two 2'-modified double-helical RNAs, poly(2'-fluoro-2'-deoxyinosinic acid) · poly(cytidylic acid) and poly(2'-chloro-2'-deoxyinosinic acid) · poly(cytidylic acid). *Eur. J. Biochem. 107*:279-288, 1980.

Greene, J. J., J. L. Alderfer, I. Tazawa, S. Tazawa, P. O. P. Ts'o, J. A. O'Malley, and W. A. Carter. Interferon induction and its dependence on the primary and secondary structure of poly(inosinic acid) · poly(cytidylic acid). *Biochemistry 17*:4214-4220, 1978.

Levine, A. S., M. Sivulich, P. H. Wiernik, and H. B. Levy. Initial clinical trials in cancer patients of polyriboinosinic-polyribocytidylic acid stabilized with poly-L-lysine, in carboxymethylcellulose [poly(ICLC)], a highly effective interferon inducer. *Cancer Res. 39*:1645-1650, 1979.

Stringfellow, D. A., and S. D. Weed. Interferon induction by and toxicity of polyriboinosinic acid [poly(rI)] · polyribocytidylic acid [poly(rC)], mismatched analog poly(rI) · poly[r(C_{12}uracil)n], and poly(rI) · poly(rC) L-lysine complexed with carboxymethylcellulose. *Antimicrob. Agents Chemother. 17*:988-992, 1980.

Mechanism of Interferon Induction

Cavalieri, R. L., E. A. Havell, J. Vilcek, and S. Pestka. Induction and decay of human fibroblast interferon mRNA. *Proc. Natl. Acad. Sci. USA 74*:4415-4419, 1977.

Meager, A., H. Graves, D. C. Burke, and D. M. Swallow. Involvement of a gene on chromosome 9 in human fibroblast interferon production. *Nature 280*:493-495, 1979.

Sehgal, P. B., and I. Tamm. Two mechanisms contribute to the superinduction of poly(I) · poly(C)-induced human fibroblast interferon production. *Virology 92*:240-244, 1979.

Sehgal, P. B., B. Dobberstein, and I. Tamm. Interferon messenger RNA content of human fibroblast during induction, shutoff, and superinduction of interferon production. *Proc. Natl. Acad. Sci. USA 74*:3409-3413, 1977.

Mass Production of Interferon

Adolf, G. R., and P. Swetly. Interferon production by human lymphoblastoid cells is stimulated by inducers of Friend cell differentiation. *Virology 99*:158-166, 1979.

Billiau, A., J. Van Damme, F. Van Leuven, V. G. Edy, M. De Ley, J-J. Cassiman, H. Van den Berghe, and P. De Somer. Human fibroblast interferon for clinical trials: Production, partial purification, and characterization. *Antimicrob. Agents Chemother. 16*:49-55, 1979.

R., E. Remaut, E. Saman, P. Stanssens, E. De Clercq, ,ontent, and W. Fiers. Expression of human fibroblast inter-ıeron gene in *Escherichia coli*. *Nature 287*:193-197, 1980.

Langford, M. P., J. A. Georgiades, G. J. Stanton, F. Dianzani, and H. M. Johnson. Large-scale production and physicochemical characterization of human immune interferon. *Infect. Immun. 26*: 36-41, 1979.

Nagata, S., H. Taira, A. Hall, L. Johnsrud, M. Streuli, J. Ecsödi, W. Boll, K. Cantell, and C. Weissmann. Synthesis in *E. coli* of a polypeptide with human leukocyte interferon activity. *Nature 284*:316-320, 1980.

Action of Interferon at the Molecular Level

Epstein, D. A., P. F. Torrence, and R. M. Friedman. Double-stranded RNA inhibits a phosphoprotein phosphatase present in interferon-treated cells. *Proc. Natl. Acad. Sci. USA 77*:107-111, 1980.

Gupta, S. L., B. Y. Rubin, and S. L. Holmes. Interferon action: Induction of specific proteins in mouse and human cells by homologous interferons. *Proc. Natl. Acad. Sci. USA 76*:4817-4821, 1979.

Hovanessian, A. G., and J. N. Wood. Anticellular and antiviral effects of pppA(2'p5'A)n. *Virology 101*:81-90, 1980.

Maheshwari, R. K., and R. M. Friedman. Effect of interferon treatment on vesicular stomatitis virus (VSV): Release of unusual particles with low infectivity. *Virology 101*:399-407, 1980.

Sen, G. C., and N. H. Sarkar. Effects of interferon on the production of murine mammary tumor virus by mammary tumor cells in culture. *Virology 102*:431-443, 1980.

Verhaegen, M., M. Divizia, P. Vandenbussche, T. Kuwata, and J. Content. Abnormal behavior of interferon-induced enzymatic activities in an interferon-resistant cell line. *Proc. Natl. Acad. Sci. USA 77*:4479-4483, 1980.

Wallach, D., and M. Revel. An interferon-induced cellular enzyme is incorporated into virions. *Nature 287*:68-70, 1980.

Williams, B. R. G., R. R. Golgher, R. E. Brown, C. S. Gilbert, and I. M. Kerr. Natural occurrence of 2-5A in interferon-treated EMC virus-infected L cells. *Nature 282*:582-586, 1979.

Action of Interferon at the Cellular Level

Ankel, H., C. Krishnamurti, F. Besancon, S. Stefanos, and E. Falcoff. Mouse fibroblast (type I) and immune (type II) interferons: Pronounced differences in affinity for gangliosides and in antiviral and antigrowth effects on mouse leukemia L-121OR cells. *Proc. Natl. Acad. Sci. USA 77*:2528-2532, 1980.

Blalock, J. E. A small fraction of cells communicates the maximal interferon sensitivity to a population. *Proc. Soc. Exp. Biol. Med. 162*:80-84, 1979.

Blalock, J. E., J. Georgiades, and H. M. Johnson. Immune-type interferon-induced transfer of viral resistance. *J. Immunol. 122*: 1018-1021, 1979.

Brouty-Boyé, D., and B. R. Zetter. Inhibition of cell motility by interferon. *Science 208*:516-518, 1980.

Gresser, I., J. De Maeyer-Guignard, M. G. Tovey, and E. De Maeyer. Electrophoretically pure mouse interferon exerts multiple biologic effects. *Proc. Natl. Acad. Sci. USA 87*:5308-5312, 1979.

Trinchieri, G., and D. Santoli. Anti-viral activity induced by culturing lymphocytes with tumor-derived or virus-transformed cells. *J. Exp. Med. 147*:1314-1333, 1978.

Zarling, J. M., L. Eskra, E. C. Borden, J. Horoszewicz, and W. A. Carter. Activation of human natural killer cells cytotoxic for human leukemia cells by purified interferon. *J. Immunol. 123*: 63-70, 1979.

Zarling, J. M., J. Sosman, L. Eskra, E. C. Borden, J. S. Horoszewicz, and W. A. Carter. Enhancement of T cell cytotoxic responses by purified human fibroblast interferon. *J. Immunol. 121*:2002-2004, 1978.

Efficacy of Interferon in Experimental Situations

Bart, R. S., N. R. Porzio, A. W. Kopf, J. T. Vilcek, E. H. Cheng, and Y. Farcet. Inhibition of growth of B16 murine malignant melanoma by exogenous interferon. *Cancer Res. 40*:614-619, 1980.

Gresser, I., C. Maury, and D. Brouty-Boyé. Mechanism of the antitumour effect of interferon in mice. *Nature 239*:167-168, 1972.

Schellekens, H., W. Weimar, K. Cantell, and L. Stitz. Antiviral effect of interferon in vivo may be mediated by the host. *Nature 278*:742, 1979

Weimar, W., L. Stitz, A. Billiau, K. Cantell, and H. Schellekens. Prevention of vaccinia lesions in rhesus monkeys by human leucocyte and fibroblast interferon. *J. Gen. Virol. 48*:25-30, 1980.

Efficacy of Interferon in Clinical Situations

Gutterman, J. U., G. R. Blumenschein, R. Alexanian, H-Y. Yap, A. U. Buzdar, F. Cabanillas, G. N. Hortobagyi, E. M. Hersh, S. L. Rasmussen, M. Harmon, M. Kramer, and S. Pestka. Leucocyte interferon induced tumor regression in human metastatic breast cancer, multiple myeloma, and malignant lymphoma. *Ann. Intern. Med. 93*:399-406, 1980.

Mellstedt, H., A. Ahre, M. Björkholm, G. Holm, B. Johansson, and H. Strander. Interferon therapy in myelomatosis. *Lancet i*:245-247, 1979.

Merigan, T. C., K. H. Rand, R. B. Pollard, P. S. Abdallah, G. W. Jordan, and R. P. Fried. Human leukocyte interferon for the treatment of herpes zoster in patients with cancer. *N. Engl. J. Med. 298*:981-987, 1978.

Merigan, T. C., K. Sikora, J. H. Breeden, R. Levy, and S. A. Rosenberg. Preliminary observations on the effect of human leukocyte interferon in non-Hodgkin's lymphoma. *N. Engl. J. Med. 299*:1449-1453, 1978.

Pazin, G. J., J. A. Armstrong, M. T. Lam, G. C. Tarr, P. J. Jannetta, and M. Ho. Prevention of reactivated herpes simplex infection by human leukocyte interferon after operation on the trigeminal root. *N. Engl. J. Med. 301*:225-230, 1979.

Weimar, W., R. A. Heijtink, F. J. P. Ten Kate, S. W. Schalm, N. Masurel, H. Schellekens, and K. Cantell. Double-blind study of leucocyte interferon administration in chronic HBsAg-positive hepatitis. *Lancet i*:336-338, 1980.

Problems Involved in Interferon Therapy

Gatmaitan, B. G., R. C. Legaspi, H. B. Levy, and A. M. Lerner. Modified polyriboinosinic acid · polyribocytidylic acid complex: Induction of serum interferon, fever, and hypotension in rabbits. *Antimicrob. Agents Chemother. 17*:49-54, 1980.

Gresser, I., L. Morel-Maroger, P. Verroust, Y. Riviére, and J. C. Guillon. Anti-interferon globulin inhibits the development of glomerulonephritis in mice infected at birth with lymphocytic choriomeningitis virus. *Proc. Natl. Acad. Sci. USA 75*:3413-3416, 1978.

Heremans, H., A. Billiau, A. Colombatti, J. Hilgers, and P. De Somer. Interferon treatment of NZB mice: Accelerated progression of autoimmune disease. *Infect. Immun. 21*:925-930, 1978.

Morel-Maroger, L., J. C. Sloper, J. Vinter, D. Woodrow, and I. Gresser. An ultrastructural study of the development of nephritis in mice treated with interferon in the neonatal period. *Lab. Invest. 39*:513-522, 1978.

Priestman, T. J. Initial evaluation of human lymphoblastoid interferon in patients with advanced malignant disease. *Lancet ii*:113-118, 1980.

chapter **4**

Type A Viral Hepatitis

JAMES E. MAYNARD and DANIEL W. BRADLEY

U.S. Public Health Service
Centers for Disease Control
Phoenix, Arizona

ETIOLOGY

Introduction

The use of the term *type A viral hepatitis* or *hepatitis A* has replaced all previous synonyms, which have included catarrhal jaundice, infectious icterus, Botkins disease, epidemic hepatitis, short incubation period hepatitis, MS-1 hepatitis, and most frequently, infectious hepatitis. Hepatitis A is a disease caused by the type A hepatitis virus, a morphologically, biochemically, and immunologically distinct agent which produces symptoms in humans after an incubation period of approximately 15 to 45 days.

Historical Perspectives

Hepatitis A was probably described by Hippocrates, who used the term *epidemic jaundice* to characterize icteric illness with a seasonal periodicity. That such jaundice was contagious was also mentioned in the eighth century A.D. in a letter from Pope Zacharias to St. Boniface urging that patients with jaundice be separated from others for fear of contagion. Extensive outbreaks of jaundice occurred in the German civilian population in 1629, in the British Army in Flanders in 1743, and at the time of the Franco-Prussian War in 1870. Sixty-three epidemics of infectious jaundice in the United States were reported between 1812 and 1920. The infection was characterized as a disease of childhood and early adult life, having an incubation period of about 28 days, spread by person-to-person contact, with greatest incidence in fall and winter months. It was later suggested that the disease might be caused by a virus. These early observations established the basis for definition of one of the forms of viral hepatitis which, as proposed in the 1940s, is now called hepatitis A.

Recovery of the Etiologic Agent

The viral etiology of hepatitis A was firmly established during World War II in human volunteer studies. In 1947, investigators at the Willowbrook State School, New York, successfully induced hepatitis in pediatric subjects who were fed bacteria-free filtrates of feces from children acutely ill with hepatitis. They showed that virus was present in feces during the incubation period of disease up to 3 weeks before onset of jaundice. In 1967, it was shown that a serum pool collected from a child just prior to onset of jaundice could reproducibly induce hepatitis in children after an incubation period of 31 to 38 days. This pool was designated as the MS-1 pool and has been extensively used in further studies as a pedigreed source of hepatitis A virus.

The hepatitis A virus (HAV) was first visualized in 1973 in the feces of human volunteers experimentally infected with the MS-1 strain of HAV. Very soon thereafter it was visualized in feces of naturally infected individuals in both the United States and Australia. Since that time the virus has been recovered in all areas in which it has been sought, and highly sensitive and specific immunologic procedures have been developed for its detection as well as the specific antibody directed against it (anti-HAV).

At the present time, only one antigenic class of HAV has been recognized which elicits a homotypic antibody response in experimental animals and in humans. This antibody is officially termed "anti-HAV." Although the existence of possible antigenic differences, as well as differences in virulence, for strains of HAV have been postulated based upon clinical and epidemiologic observations, specific laboratory evidence for such differences is currently lacking.

Recently, the successful propagation of HAV in cell culture has been reported, although conditions for optimal viral replication have yet to be established. The immediate future promises success in the development of a hepatitis A vaccine from virus grown in cell culture.

BIOCHEMICAL AND BIOPHYSICAL PROPERTIES OF HAV

Morphology

Hepatitis A virus is a 27 nm spherical, nonenveloped particle with a cubic icosahedral symmetry. Platinum-palladium metal shadowing of partially purified particles (Fig. 1) clearly reveals the spherical geometry of the virus. Both empty and full virus particles can be visualized

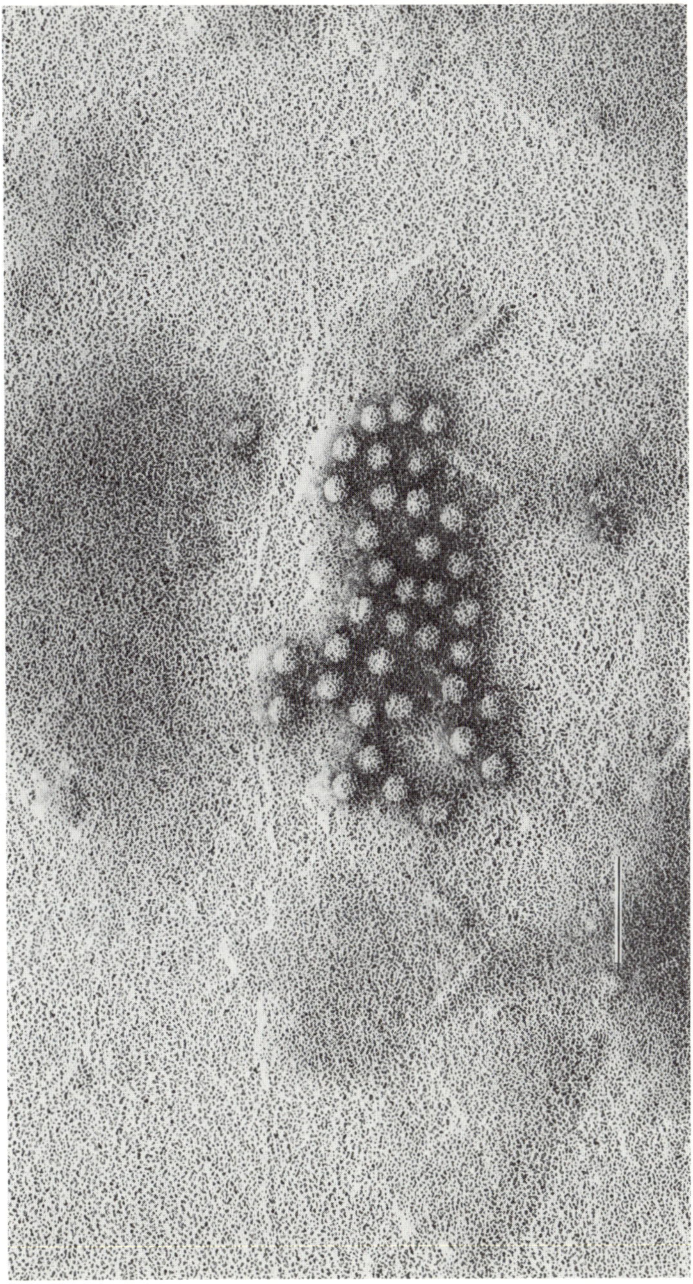

Figure 1 Electron micrograph of 27 nm hepatitis A virus particles negatively stained with phosphotungstic acid (PTA) and metal shadowed with platinum-palladium. Magnification 137,600X. (From Department of Health and Human Services, Hepatitis Laboratories Division, Phoenix, Arizona.)

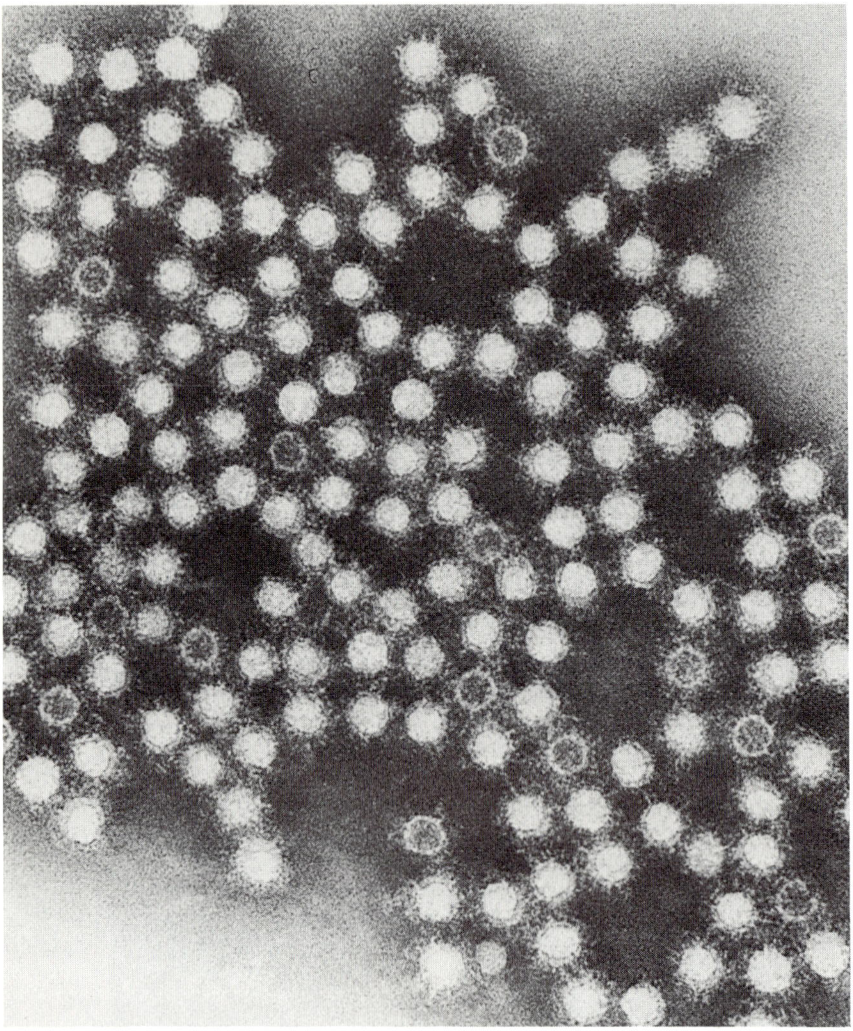

Figure 2 Electron micrograph of 27 nm HAV particles aggregated by antibody to HAV and negatively stained with PTA. Magnification 257,200X. (From Department of Health and Human Services, Hepatitis Laboratories Division, Phoenix, Arizona.)

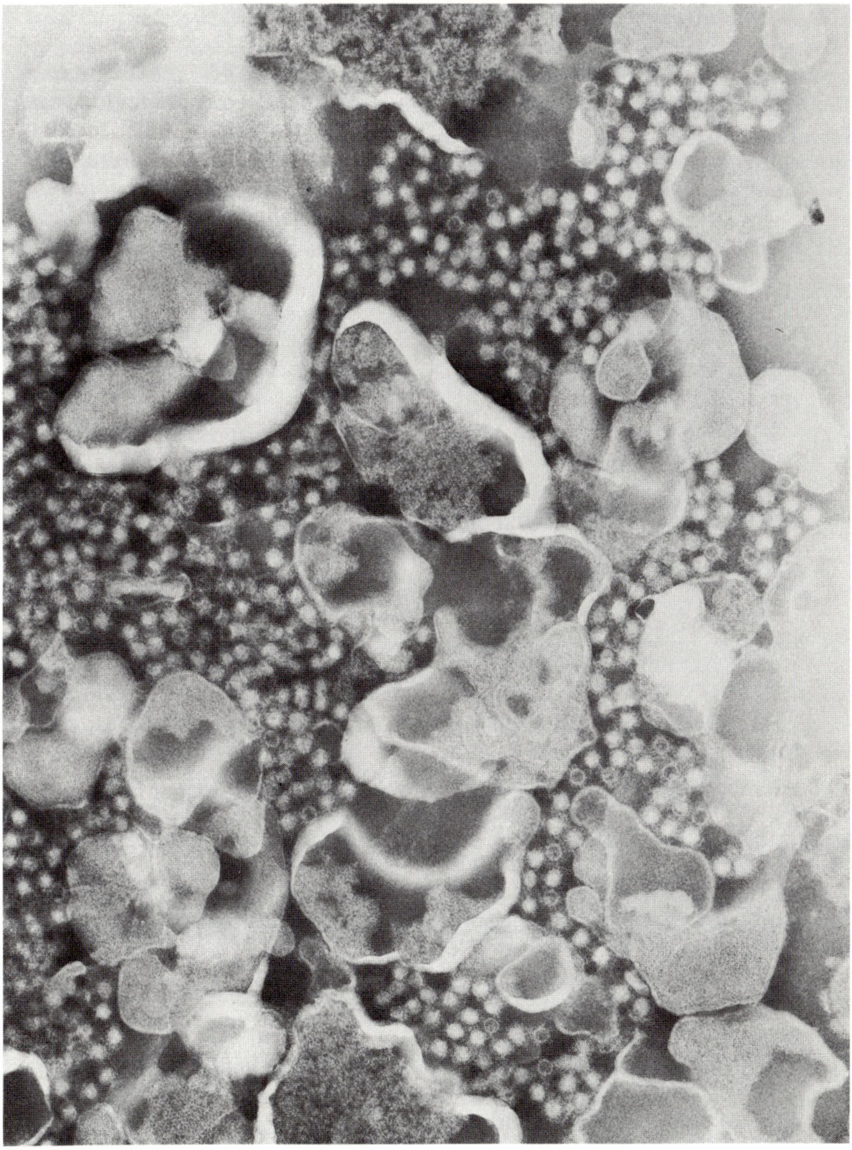

Figure 3 Electron micrograph of HAV particles recovered from an acute-phase marmoset liver banded on a cesium chloride density gradient (buoyant density = 1.34 g/cm^3). Magnification 88,500X. (From Department of Health and Human Services, Hepatitis Laboratories Division, Phoenix, Arizona.)

by immune electron microscopy (IEM) in infected liver, bile, and feces. Full HAV particles contain an electron-dense ribonucleic acid genome that often gives the appearance of a "corelike" structure, while empty particles, that is, protein capsids devoid of nucleic acid, resemble hollow shells (Fig. 2). Virus recovered from acute-phase chimpanzee or marmoset liver homogenates is often found associated with cellular debris or membranous structures (Fig. 3).

Stability

Purified HAV is inactivated by treatment with formalin (0.025% w/v formaldehyde, 72 h at 37°C), ultraviolet irradiation (1.1 W for 60 s), heating at 100°C for 5 min, or treatment with chlorine (1 ppm for 30 min). HAV is partially inactivated by heating at 60°C for 1 h and by treatment with pancreatic ribonuclease. Extraction of HAV from infected feces or acute-phase liver homogenates using nonionic detergents (Triton X-100, Nonidet P-40), chloroform, ether, or fluorocarbon (1,1,2-trichlorotrifluoroethane) does not result in activation or alteration of virus particle morphology. HAV is also stable at pH 3.0 for 3 h at room temperature. These properties of HAV are consistent with the fact that infectious virus can be readily recovered from bile and feces (Fig. 4) and that lipids or lipoproteins are not structural components of the virus capsid. Although feces containing HAV can be stored for long periods of time (more than 10 years at -50°C) without complete loss of infectivity, freeze-thawing of infected liver homogenates or feces may result in a significant loss of virus particles, as determined by electron microscopy. The recovery of intact HAV from liver homogenates kept at -70°C has also been shown to be influenced by the length of time in storage and may reflect gradual denaturation of capsid protein. Treatment of HAV-positive liver homogenates with 0.2% w/v formaldehyde may enhance the recovery of intact virus particles, presumably by stabilization of the capsid structure.

Biophysical Properties

Buoyant Density

Preliminary studies conducted in 1973 showed that HAV recovered from the feces of an experimentally infected volunteer banded in a cesium chloride (CsCl) density gradient at a buoyant density of approximately 1.40 g/cm^3. HAV was postulated to be a parvovirus, based on its size, buoyant density, and known stability to acid, ether, and heat (56°C for 30 min) treatment. Additional investigations completed in 1974, however, demonstrated that HAV infectivity contained in the acute-phase serum of an experimentally infected marmoset banded in a CsCl gradient primarily at a buoyant density of 1.34 g/cm^3.

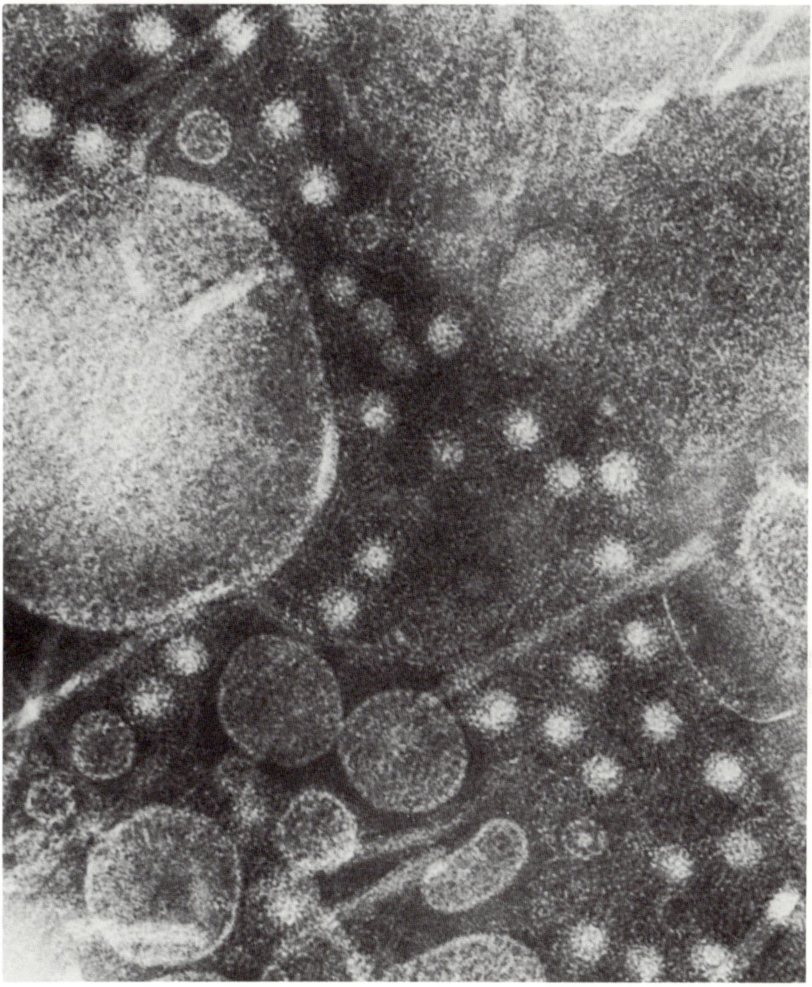

Figure 4 Electron micrograph of 27 nm HAV particles recovered from acute-phase human feces. Virus particles are heavily coated by anti-HAV antibody used in the IEM assay. Magnification 256,000X. (From Department of Health and Human Services, Hepatitis Laboratories Division, Phoenix, Arizona.)

A lesser peak of infectivity was found at a buoyant density of 1.15 g/cm^3, suggesting that virus was associated with lipid-containing structures, i.e., membranes. This same study also revealed the presence of 27 nm diameter virus-like particles in thin-sectioned acute-phase marmoset liver. The virus was found in the hepatocyte cytoplasm, but not in the nucleus, and tended to be localized in vesicles bounded by multilayered membranes. Other workers concurrently showed that HAV recovered from acute-illness phase human and chimpanzee feces banded primarily at 1.33 g/cm^3 in CsCl density gradients. Minor peaks of HAV particles were detected by immune electron microscopy at buoyant densities of 1.41 and 1.50 g/cm^3. These combined findings are most consistent with the concept that HAV is an enterovirus and not a parvovirus. Subsequent studies further documented the existence of HAV in feces with multiple buoyant density properties in CsCl gradients. Major virus peaks were found at densities as low as 1.29 g/cm^3 and as high as 1.48 g/cm^3. HAV particles and antigen could also be detected at 1.23 g/cm^3, as well as at buoyant densities ranging between 1.29 and 1.48 g/cm^3, with a predominant peak at 1.34 g/cm^3. Virus banding at 1.34 and 1.45 g/cm^3 was shown to be morphologically similar and antigenically indistinguishable, suggesting the 1.45 g/cm^3 particle was a structural variant of the 1.34 g/cm^3 archetype. HAV particles detected by electron microscopy at a buoyant density of 1.30 g/cm^3 or less appeared to be empty or incomplete, whereas virions banding at higher densities were primarily full. CsCl density gradient fractions containing HAV antigen banding at buoyant densities of 1.23 g/cm^3 or less were often found by electron microscopy to be negative for HAV particles. HAV antigen banding at such low buoyant densities was presumed to be soluble protein.

The complexity of HAV banding profiles in CsCl density gradients was unexpected; however, recognition of the multiplicity of virus buoyant densities helped explain the earlier discrepancies in its reported banding properties. The question of whether these buoyant densities of fecal HAV were artifacts of analysis or were inherent properties of the virus was answered in part by analysis of serial (daily) fecal specimens obtained from experimentally infected chimpanzees. Isopycnic banding of HAV-positive preacute-phase feces revealed empty HAV particles banding primarily at a buoyant density of 1.29 to 1.30 g/cm^3 in CsCl. Analysis of HAV in feces collected 5 days later, but still before alanine aminotransferase activity (ALT) became elevated, showed a bimodal banding pattern of HAV. A minor peak of HAV particles was found at a buoyant density of 1.29 g/cm^3; a major HAV peak was detected by IEM and radioimmunoassay at a buoyant density of 1.33 g/cm^3. Isopycnic banding in CsCl of HAV in feces collected during the initial elevation of ALT activity also revealed a bimodal distribution of virus particles. However, the major and minor HAV peaks were found at buoyant densities of 1.33 and 1.40 g/cm^3, respectively. These findings suggested that the observed multiplicity of HAV buoyant densities in CsCl was temporally related to the course of disease. In addition, the recovery of empty capsids from early feces is consistent with the observation that enterovirus assembly in vivo involves the formation of procapsids (empty virus) from capsomeric subunits, followed by the envelopment of nascent viral RNA. The appearance of HAV with abnormally high buoyant densities during the later phase of disease may reflect the production of defective virions. For example, enterovirus particles with buoyant densities of > 1.40 g/cm^3 have been shown to be less infectious than particles banding at 1.34 g/cm^3.

Sedimentation Coefficient

Additional studies of the light (1.34 g/cm^3) and heavy (1.45 g/cm^3) HAV particles described above indicated that they were structurally different. Rate-zonal banding of light- and heavy-density HAV in neutral pH sucrose gradients showed that both particle types had sedimentation constants of 157S. However, when light and heavy HAV were banded in sucrose gradients containing 0.5, 1.0, or 1.5 M CsCl, heavy particles were found to sediment faster than light particles. Increased concentrations of CsCl led to an increased sedimentation coefficient of the heavy HAV when compared to that of the light-density virus. These properties of light and heavy HAV are similar to those reported for a variety of enteroviruses, most of which have sedimentation coefficients of around 160S in neutral pH sucrose gradients.

The increased sedimentation coefficient of heavy HAV particles in sucrose gradients containing CsCl was postulated to be a result of increased binding of cesium ions to viral nucleic acid. Penetration of loosely structured "heavy" HAV capsids by Cs^+ would result in significant increases in both the particle density and its sedimentation coefficient. The observed behavior of heavy HAV in sucrose gradients without CsCl also suggests that these particles have the same nucleic acid content as the morphologically identical light-density HAV. Rate-zonal banding of empty HAV particles, that is, virus banding in CsCl at 1.30 g/cm^3, shows that they have sedimentation coefficients of between 50S and 93S. HAV antigen, presumably soluble protein or capsid subunits, has been shown to have an even lower sedimentation coefficient of approximately 5S to 15S.

Nucleic Acid

The visualization of 27 nm virus-like particles in the cytoplasm of hepatocytes from infected marmosets, the reported sensitivity of HAV to ribonuclease, and the results of acridine orange staining of infected hepatocytes suggested that HAV contained a single-stranded RNA genome and was a picornavirus. More recent studies have provided substantial evidence for the notion that HAV should be further classified as an enterovirus. Treatment of purified 1.34 g/cm^3 HAV (160S particles) at pH 12.7 resulted in the release and subsequent hydrolysis of the viral genome. Any remaining nucleic acid found by electron microscopy was present in short, linear, single-stranded molecules approximately 0.2 to 0.6 μm in length. These results suggested the viral genome was comprised of single-stranded RNA, since it is known that double-stranded RNAs and single-stranded DNAs are relatively insensitive to alkaline pH treatment. Furthermore, alkaline pH treatment to single-stranded RNAs obtained from a variety of picornaviruses has previously been shown to result in their hydrolysis.

Other studies have shown that mild alkaline pH treatment of light (1.34) and heavy (1.45) HAV particles resulted in the formation of labile structures that were sensitive to ribonuclease, but not deoxyribonuclease, digestion. In particular, HAV particles with a buoyant density of 1.34 g/cm^3 and a sedimentation coefficient of 157S were shown to be disrupted by pH 10.0 treatment, since approximately one-half the particles and HAV antigen sedimenting at 157S was lost after treatment. The remaining virus particles were found to be exquisitely sensitive to digestion by pancreatic ribonuclease, even at enzyme concentrations as low as 10^{-4} μg/ml. RNase digestion of pH 9.0-treated heavy HAV particles also resulted in the loss of the majority of particles and HAV antigen banding at 157S in sucrose gradients. DNase treatment of pH 10.0-treated light-density virus did not result in a loss of HAV particles or antigen banding at 157S. These results also suggested that the genome of HAV was a single-stranded RNA molecule, since pancreatic RNase A (EC 3.1.4.22) is known to hydrolyze only single- and not double-stranded RNA.

Polypeptide Composition

Analysis of purified HAV (1.34 g/cm^3, derived from feces) by discontinuous SDS-PAGE (sodium dodecyl sulfate-polyacrylamide gel electrophoresis) has revealed four major polypeptides with molecular weights similar to the four polypeptides of poliovirus. SDS-PAGE of ^{125}I-labeled purified HAV showed polypeptides with molecular weights of 29,500 daltons, 24,000 d, 22,000 d, and 14,00 d. SDS-PAGE of purified HAV followed by Coomassie blue staining of the protein bands revealed polypeptides with molecular weights of 34,000 d, 25,500 d, and 23,000 d. These latter molecular weights are in remarkably close agreement with the accepted molecular weights of poliovirus VPI (33,000 d), VP2 (23,000 d), and VP3 (23,000 d).

Purification

HAV has been purified by a variety of techniques for (1) use as a reagent in serologic tests for antibody to HAV (anti-HAV), (2) analysis of its nucleic acid genome, (3) analysis of its polypeptide composition, and (4) use in tissue culture. Purification techniques have included affinity chromatography, molecular sieve chromatography, ion-exchange chromatography, rate-zonal banding in sucrose gradients, isopycnic banding in CsCl density gradients, isoelectric focusing, precipitation with polyethylene glycol (PEG), and extraction with nonionic detergents, chloroform, ether, or fluorocarbon.

HAV with a density of 1.33 g/cm^3 has been successfully purified from preacute and acute-phase chimpanzee feces by a combination of PEG precipitation, CsCl banding, and molecular sieve chromatography. These three procedures take advantage of the fact that HAV antigen is a high molecular weight nucleoprotein, possesses a buoyant density significantly different from the bulk of the stool materials, and has a diameter that excludes it from more than 99% of the materials in feces when chromatographed on Sepharose CL-2B. Virus purified by these techniques was shown by electron microscopy to be free of contaminating fecal detritus (Fig. 5). The purified virus was also shown by IEM to be specifically aggregated by anti-HAV. Inoculation of purified HAV into a susceptible chimpanzee induced enzymatic, serologic, and histopathologic evidence compatible with a diagnosis of acute viral hepatitis A. This finding proved the infectivity of the highly purified HAV and further established its identity with the agent responsible for hepatitis A. An alternate procedure for the purification of HAV from feces employed differential centrifugation, organic solvent extraction, agarose gel filtration, ion-exchange chromatography, and isopycnic ultracentrifugation in CsCl. HAV purified by this method was shown to be suitable for polypeptide analysis.

Classification of Virus

The morphologic dimensions of the 27 nm HAV particle, its polypeptide composition, buoyant density in CsCl, sedimentation coefficient in sucrose, genome of single-stranded RNA, and acid, ether, and heat stability are most consistent with its classification as an enterovirus within the family Picornaviridae. Although HAV is more heat stable than some other enteroviruses, and possesses a lower molecular weight RNA than other enteroviruses (1.9×10^6 d versus 2.5×10^6 d), the bulk of its biochemical and biophysical properties favor its classification as an enterovirus.

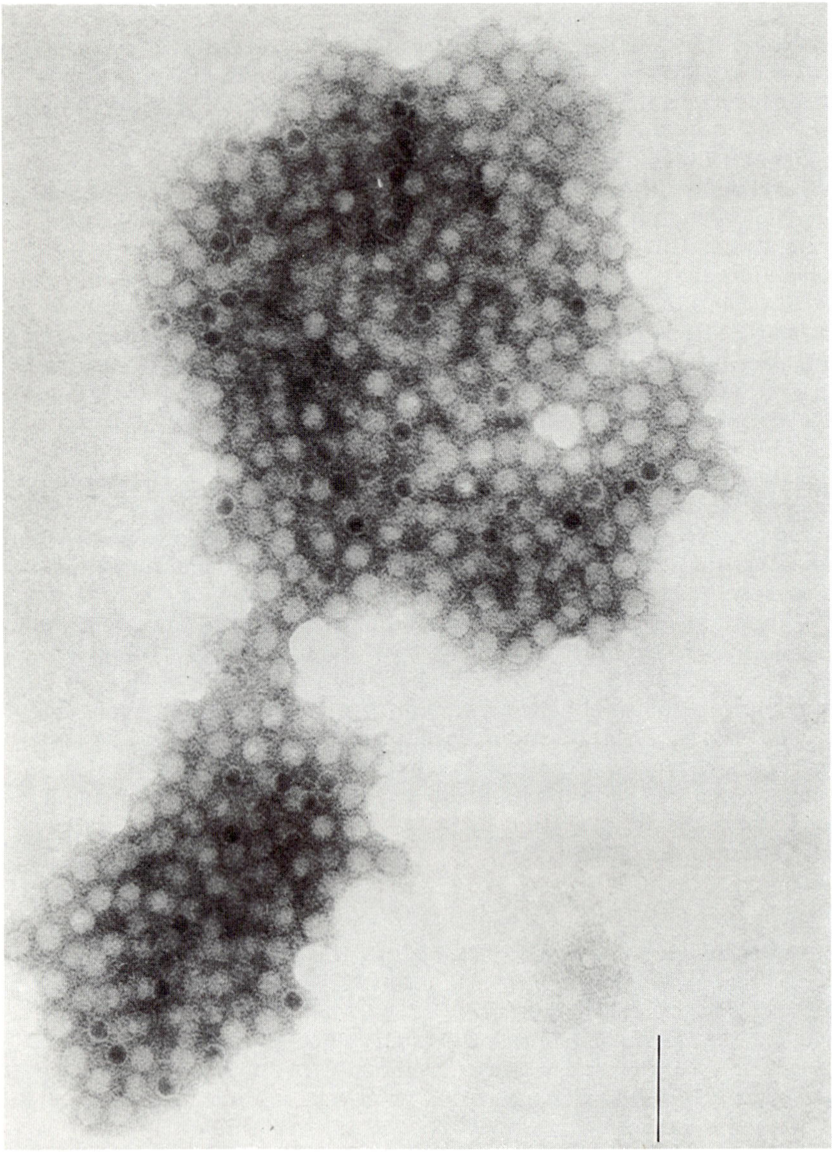

Figure 5 Electron micrograph of HAV purified from acute-phase chimpanzee feces. Particles are aggregated by anti-HAV antibody used in the IEM assay. Magnification 137,600. (From D. W. Bradley et al., *Journal of Virological Methods* 2:35, 1980.)

EPIDEMIOLOGY

Hepatitis A is a disease of worldwide distribution which occurs with both endemic and epidemic patterns. Classically it has affected mainly children, with a demonstrable inverse effect of socioeconomic status and sanitary level on age-specific incidence of infection. In areas of high prevalence with conditions of poor environmental sanitation, infection occurs in early childhood, with as high as 95% of the population having experienced infection by the age of 10 to 12 years. These conditions exist in many developing countries of Asia and Africa. Here, the clinical disease in small children is usually mild and the incidence of asymptomatic infection is high. In areas of higher socioeconomic level, such as Western Europe and North America, where good sanitation prevails, environmental barriers and increased host resistance may prevent spread of infection in early childhood. Here, occurrence of disease is delayed, often to young or middle adulthood, and epidemics are observed where susceptible individuals are aggregated in special circumstances, such as institutions for the mentally retarded, military barracks, colleges, and summer camps. A high frequency of hepatitis may also be observed in residents of areas of low endemicity who travel for business or pleasure to areas of high endemicity. Clinical disease in adults tends to be more severe than in children and the relative frequency of subclinical infection lower.

Despite the high endemicity of HAV infection in India and Southeast Asia, large-scale outbreaks of hepatitis have been observed in these areas on several occasions with a high incidence of illness in adults and high mortality rates in pregnant women. The reasons for the occurrence of such epidemics are not yet well understood, although several explanations have been postulated, including high dose of virus in the face of waning immunity and hypothetical differences in the strain characteristics of HAV.

The efficient worldwide penetration of HAV is illustrated by the following age-standardized prevalences (in percentages) for anti-HAV in serum determined for several countries as follows: USA, 44.7; Switzerland, 28.7; Belgium, 81.1; Yugoslavia, 96.9; Israel, 95.3; Taiwan, 88.7; and Senegal, 76.2. However, cohort analysis of age-specific anti-HAV prevalences in several Western European countries indicate that HAV may be disappearing as an endemic infection, particularly since World War II, with the bulk of disease in these countries occurring in individuals returning from endemic areas.

Analysis of morbidity trends from several countries shows that hepatitis A has tended to occur in a cyclic fashion with minor yearly peaks in late summer and fall seasons and major epidemic waves every 5 to 20 years. No seasonal peaks are generally discernible in tropical areas. Recent data from the United States indicate that the seasonality of hepatitis A has largely disappeared and that its overall incidence is slowly declining.

The primary reservoir of hepatitis A is the human, although one species of old world ape, the chimpanzee, and several subspecies of new world marmoset monkeys are susceptible to infection. In regard to nonhuman primates, these animals may be experimentally infected with HAV, but there is no evidence of a sylvatic cycle of transmission. Except for transmission of hepatitis A from nonhuman primates in captivity to their handlers, transmission of HAV from nonhuman primates to humans under natural conditions has not been demonstrated.

The most common route of transmission of hepatitis A is by person-to-person contact, usually from an individual in the incubation period of disease to a susceptible, and by the fecal-oral route. HAV is presumably maintained in human populations by serial transmission of virus during the incubation period of disease. Long-term chronic fecal shedders of virus have not been shown to exist, and current evidence does not favor the existence of a respiratory mode of spread. Although the existence of viremia prior to the onset of illness has been demonstrated in experimental studies in nonhuman primates and humans, the period of viremia is short and the titer of virus appears to be low. A chronic blood carrier state for HAV does not, in all likelihood, exist, and HAV is not a cause of posttransfusion hepatitis.

In addition to spread by person-to-person contact, common-source epidemics of hepatitis A due to contamination of water and milk have been described. One of the largest urban outbreaks in modern history occurred in 1956 in Delhi, India, with over 25,000 cases attributed largely to contamination of water supplies. More common than outbreaks due to water or milk are those associated with contaminated food. A wide variety of food items, contaminated by foodhandlers in the incubation period of disease, have been incriminated, including bakery goods, salads, fish, and cold meats. Only uncooked items or cooked items handled subsequently to cooking by implicated handlers are regarded as effective vehicles of transmission. Ingestion of raw shellfish, in particular raw clams and oysters, harvested from waters contaminated by human sewage, has been responsible for outbreaks of disease.

Although it has been widely presumed that hepatitis A constitutes a nosocomial hazard which requires special isolation of acutely ill individuals while in hospital, recent serologic studies indicate that hospital personnel are at no greater risk of infection than individuals in the general community. Experimental studies also indicate that the bulk of HAV excretion also occurs prior to the occurrence of frank jaundice. Hepatitis A is, therefore, not a significant nosocomial infection, and only simple precautions in the handling of feces are required for containment.

PATHOGENESIS, COURSE OF DISEASE, AND IMMUNE RESPONSE

Pathogenesis

Infection is initiated by oral ingestion of virus, and although infection has been induced experimentally by inoculation of serum or plasma which contains HAV, this route of transmission does not appear to be significant under natural conditions. Infected hepatocytes appear to be the site of replication of HAV, but the means by which virus gains access to these liver cells is not precisely known. The ability to detect HAV viral antigen in the cytoplasm of liver cells by immunofluorescent techniques has provided a mechanism for determination of virus localization in experimental studies in marmosets and chimpanzees. To date, examination of tissues from a variety of organs, including upper and lower intestinal tract, in both marmosets and chimpanzees, has failed to yield evidence of HAV replication in any other cells except hepatocytes.

The characteristic histopathologic changes produced in hepatitis A in humans involve focal necrosis of hepatocytes with accompanying widespread infiltrations of inflammatory cells. Evidence of cellular regeneration with increased mitotic figures is concurrently present. In the typical case, the hepatocyte necrosis is scattered and cells may either break into fragments or shrink. Complete regeneration of liver cells usually occurs within 2 to 3 months of the onset of illness. In fulminant hepatitis the size of the liver becomes reduced with changes typical of acute yellow atrophy. In the lethal form, hepatocytes are destroyed with no evidence of cellular regeneration. Under conditions of survival, cellular regeneration is irregular and leads to distortion of normal liver architecture with formation of regeneration nodules surrounded by zones of collapsed reticulum-containing fibrous tissue.

Course of Disease

Clinical Manifestations

Hepatitis A has been described to have a relatively abrupt onset with development of signs and symptoms over a 24 to 72 h period. The initial symptoms are fatigue and anorexia, usually accompanied by low-grade fever and some abdominal pain or discomfort in the epigastrium or right upper quadrant. In those individuals who develop frank jaundice, the urine becomes brownish, and scleral icterus develops within 3 to 5 days. In these cases, serum bilirubin is usually above 4 mg per 100 ml. The liver may be palpable at one to two fingerbreadths below the right costal margin and is usually tender. In mild and anicteric cases, these signs and symptoms may be inconstant, and the diagnosis confirmed only by the presence of abnormal biochemical tests or the presence of bile or bile products in the urine.

Prognosis

The clinical course of hepatitis A is usually self-limited, with full recovery after a period of 2 to 4 weeks. Chronic liver sequellae occur extremely rarely, if at all. In a few instances, estimated from two recent studies to be approximately 0.3% of hospitalized cases, a severe fulminant hepatitis develops. Approximately 78 to 80% of these cases die.

Treatment

There is no specific treatment for benign, acute hepatitis A. Therapy should be supportive and aimed at maintaining comfort and adequate nutritional balance. Because of the mildness of the disease, particularly in children, bed rest is not required beyond the acute phase of illness and should be tailored to the patient's own inclination to remain in bed. In fact, controversy exists regarding the necessity for any bed rest. Two studies have failed to show any adverse effect of continued ambulation on the clinical course of acute, viral hepatitis in otherwise healthy individuals when compared with control patients put on bed rest. Prudent therapy, however, seems to indicate a course of bed rest carefully tailored to the patient's illness and mental well-being without insistence on prolonged inactivity. Hospitalization is indicated only in severe cases or in circumstances in which adequate care cannot be maintained in the home environment.

The diet should be tailored to the patient's appetite and wishes; it must, however, contain adequate protein and calories. There is no good evidence that restriction of fats has any beneficial effect on the course of illness, and ingestion of milk products in moderation may help to assure adequate caloric intake. In patients hospitalized for severe anorexia or nausea and vomiting, intravenous therapy to assure adequate electrolyte balance and to meet basal caloric requirements should be instituted.

There is no indication for the use of adrenocortical steroids in the treatment of acute, uncomplicated viral hepatitis, and recent controlled studies show no beneficial effect of such treatment in management of severe, acute viral hepatitis, including fulminant hepatitis.

The treatment of acute, fulminant hepatitis has been unsatisfactory; and case fatality, as previously indicated, remains high. Conventional treatment has included complete elimination of protein intake and introduction of oral neomycin therapy, use of parenteral vitamin K therapy to control bleeding, and intravenous administration of glucose and enemas to empty the intestinal tract. The more heroic procedures, including exchange-transfusion, extracorporeal liver perfusion, plasmapheresis with plasma exchange, and cross-circulation, in humans and nonhuman primates, have not shown conclusive increased benefit over more conservative management efforts.

Immune Response

Infection with HAV typically results in the appearance of clinical symptoms 15 to 45 days later (Fig. 6). HAV particles and antigen may be excreted in feces as early as 1 week after inoculation, but are often absent in feces collected during the acute phase of illness. The bulk of virus is usually shed before the onset of clinical symptoms. Anti-HAV can be detected by IEM and radioimmunoassay (RIA) in acute-illness phase sera as well as in convalescent-phase sera. As in other virus diseases, infection with HAV results in an early antibody response primarily of the IgM class of immunoglobulins. Acute-illness phase IgM anti-HAV may account for more than 95% of the anti-HAV present in serum or plasma. For example, fractionation of preacute-phase plasma from an experimentally infected chimpanzee showed that IgG anti-HAV constituted less than 1% of the total anti-HAV; the titer of IgM anti-HAV was more than 100 times greater than that of IgG anti-HAV. IgG anti-HAV is generally in low titer during the acute phase of disease; however, substantial increases in the titer of IgG anti-HAV do occur during the convalescent phase of disease. The titer of IgM anti-HAV reaches a peak during the acute phase of disease and decreases during the convalescent phase of disease. IgM

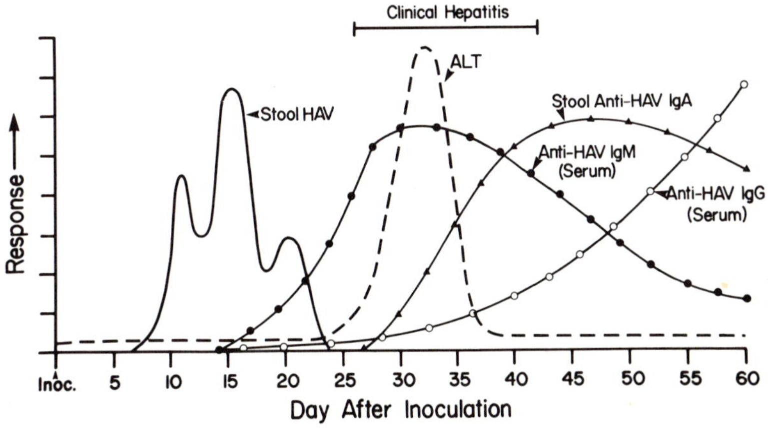

Figure 6 Typical response to infection with HAV. (From D. W. Bradley et al., *Journal of Virological Methods* 2:35, 1980.)

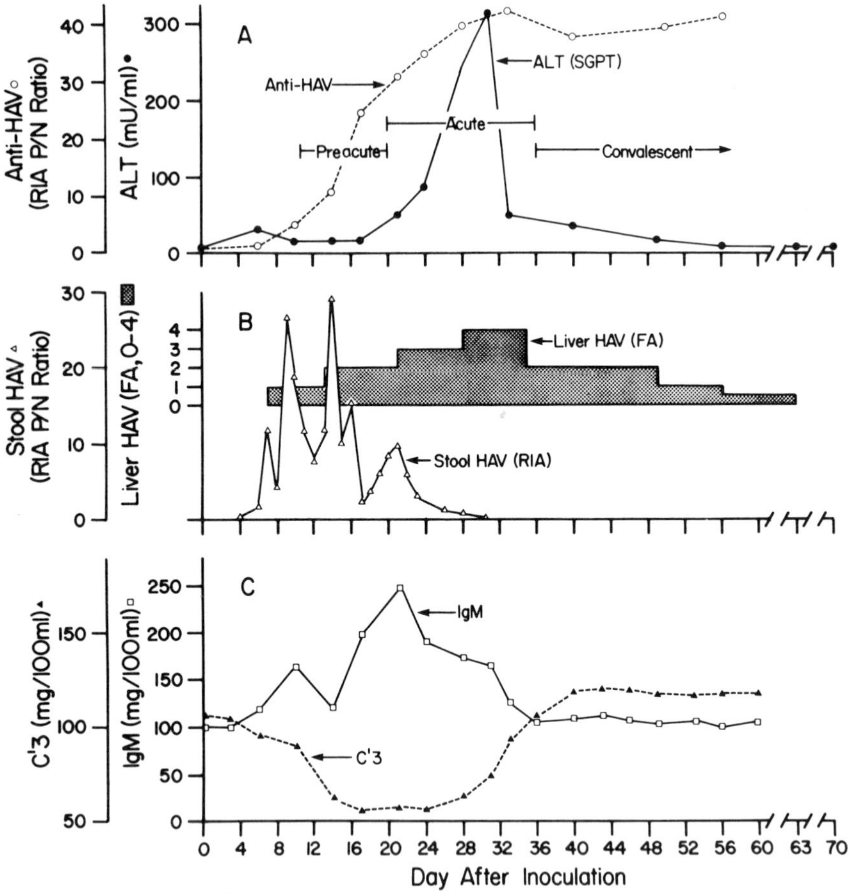

Figure 7 Response of a chimpanzee to experimental infection with HAV. (From D. W. Bradley, *Journal of Virological Methods 2*:35, 1980.)

anti-HAV usually declines to undetectable levels during the first 3 months of convalescence. Some workers have recently reported the detection of an IgA anti-HAV coproantibody in acutely infected humans; however, not all acutely infected individuals tested had demonstrable IgA anti-HAV in their feces. Although the temporal relationship of IgA anti-HAV to the appearance of clinical symptoms, elevated ALT activity, and presence of serum IgM anti-HAV has not yet been fully described, the available data suggest that the presence of IgA anti-

HAV in feces is variable. Thus, it would presently appear that tests for IgA anti-HAV in feces may not be useful for the diagnosis of acute hepatitis A infection.

The actual course of disease in an experimentally infected chimpanzee is shown in Fig. 7. Intravenous inoculation of an acute-phase stool suspension in this animal resulted in elevated ALT activity within approximately 3 weeks. ALT activity rose to a peak value 30 days after inoculation and rapidly declined to baseline values a few days later (Fig. 7A). It is interesting to note that antibody to HAV was detected in serum as early as 10 days after inoculation. HAV antigen was detected in feces and liver biopsies as early as 6 days after inoculation. HAV was maximally excreted in feces between 9 and 15 days after inoculation, but was undetectable in feces collected during the late acute phase of disease (Fig. 7B). The concentration of HAV antigen in liver was at a maximum level during the acute phase of disease and slowly declined to undetectable levels 9 weeks after inoculation. The persistence of virus or viral antigen in liver is probably related to the gradual clearance of HAV from infected hepatocytes. The absence of detectable HAV antigen in feces at a time when liver is still positive for virus may reflect the formation of antigen-antibody complexes comprised of IgA anti-HAV or IgM anti-HAV and HAV antigen. Virus or viral antigen in such complexes in feces may be undetectable by IEM or RIA. The appearance of anti-HAV in serum is temporally associated with increased concentrations of total IgM and decreased serum levels of complement component C'3 (Fig. 7C). A significant proportion of the total IgM has been found to be comprised of IgM anti-HAV. The consumption of C'3 during the time when IgM anti-HAV and total IgM are at their highest levels in serum is thought to be related to the formation of immune complexes, some of which are probably composed of IgM anti-HAV and HAV. Viremia, or the presence of virus in serum, has been shown to occur during the acute phase of disease in infected humans and chimpanzees; therefore, it seems reasonable to assume that circulating virus would combine with IgM anti-HAV and C'3 to form circulating immune complexes (CICs). The role of these CICs in the pathogenesis of hepatitis A has not been experimentally defined; however, it is interesting to note that maximum levels of IgM, minimum concentrations of C'3, and maximum shedding of virus in feces occur shortly before the peak of ALT activity.

TRANSMISSION OF HEPATITIS TO NONHUMAN PRIMATES

Initial attempts to experimentally transmit human hepatitis A to a wide variety of domestic and laboratory animals resulted in failure. However, in 1938 the reported transmission of a "hepatitis-like disease" to humans following intravenous inoculation of a yellow fever vaccine prepared from a rhesus monkey hyperimmune serum pool suggested these primates may be susceptible to human hepatitis A. The occurrence of viral hepatitis in chimpanzees (*Pan troglodytes*) was documented as early as 1963, when it was reported that biochemical and histopathologic abnormalities resembling those seen in human viral hepatitis were observed in animals which had been imported to this country. Earlier evidence that chimpanzees were involved in the transmission of hepatitis to their handlers had accumulated when outbreaks of infectious hepatitis between 1958 and 1960 were documented among chimpanzee handlers at a United States Air Force base. Additional outbreaks of infectious hepatitis were subsequently reported among handlers of imported young chimpanzees, gorillas, Celebes apes, and wooly monkeys. In more recent years, laboratory studies have unequivocally demonstrated the susceptibility of chimpanzees to human hepatitis A (see discussion below).

The development of a biochemically and histopathologically typical viral hepatitis in some species of marmosets, small South American tamarins (*Saguinus fusicollis*, *S. nigricollis*, *S.* [*Oedipomidas*] *oedipus*, and *S. mystax*), inoculated with a variety of human hepatitis materials was reported in 1965. The disease could be subpassaged in these animals by intravenous inoculation of acute-phase serum obtained from previously inoculated marmosets when elevations of serum isocitrate dehydrogenase (SICD) and liver histopathology were first seen. Although it was argued by some that the disease being passaged in these animals was caused by a reactivated latent marmoset agent, later studies showed that the etiologic agent was indeed human HAV. Subsequent reports showed that a well-characterized strain of HAV, MS-1, could also be transmitted to marmosets (*S. fusicollis*, *S. nigricollis*, *S. oedipus*) and that the disease caused by this strain of HAV could be passaged in these animals by intravenous inoculation of acute-phase serum. Another strain of HAV, CR326, was also shown to infect marmosets. Hepatitis A was successfully transmitted to 8 of 10 *S. mystax* and 4 of 12 *S. nigricollis* after intravenous inoculation of the CR326 agent. Hepatitis A has also been transmitted to *S. nigricollis*, *S. labiatus*, *S. oedipus*, and *S. mystax* marmosets by intravenous or oral inoculation of a third strain of HAV, referred to as the Phoenix antigen. This latter strain of HAV was recovered from a pool of acute-phase feces collected during an outbreak of hepatitis at a local university. It is interesting to note that serologic studies have shown that the MS-1, CR326, and Phoenix strains of HAV are antigenically similar, if not identical. Additional studies in marmosets have shown that another species similar to *S. labiatus*, designated *Marikana labiata* (*Jacchus rufiventer*), can be infected by the CR326 strain (and presumably the MS-1 and Phoenix strains) of HAV. The *J. rufiventer* species of marmoset appears to be as susceptible as *S. mystax* to infection with HAV. Several other species of marmosets, including *Callithrix argenata*, *C. jacchus*, *S. oedipus*, and *S. weddelli* have been shown to seroconvert to HAV following intravenous inoculation of the CR326 agent; however, they do not demonstrate enzymatic or histopathologic evidence of acute infection.

Unfortunately, neither chimpanzees nor marmosets are readily available in sufficient numbers for widespread use in biomedical research. Chimpanzees have been designated a threatened species and are obtained only at great cost from breeding colonies located within the United States. Supplies of marmosets, primarily imported from South America, are subject to frequent embargoes of unpredictable duration. The limited host range of HAV in animals that are most difficult to obtain has prompted renewed efforts to identify other species of susceptible nonhuman primates. The South African lesser bushbaby (*Galago senegalensis*), for example, has been shown to be susceptible to infection with HAV; however, biochemical, histopathologic, and serologic evidence of acute infection has been variable in these animals. Recent serologic surveys of new world and old world monkeys have revealed the presence of anti-HAV in a large variety of species, and seroconversions were documented in some, including rhesus monkeys inoculated with HAV. None of these species demonstrated biochemical or histopathologic evidence of acute infection. However, more intensive transmission studies in these animals may reveal limited susceptibility in some.

Marmoset Model

Intravenous or oral inoculation of certain species of marmosets with HAV results in enzymatic, histopathologic, and serologic evidence of acute hepatitis A. Unlike humans, infected marmosets do not develop frank icterus nor do they exhibit obvious clinical symptoms other than lethargy, loss of appetite, and moderate fever. Histologic changes observed in liver biopsies are often mild during the acute phase of disease and generally consist of diffuse hepatocellular necrosis, cytoplasmic vacuolization, hepatic perivascular portal inflammation, and parenchymal cell involvement. The degree of liver damage is most closely correlated with elevations in SICD activity, and less so with elevations in ALT and aspartate aminotransferase (AST) activities.

Oral or intragastric inoculation of marmosets with HAV-positive human or chimpanzee feces (or sera) is less efficient in transmitting disease than is intravenous inoculation of identical materials. For example, oral inoculation of HAV-positive human feces in *S. mystax* marmosets has been shown to cause disease in only 30% of the animals. Intravenous inoculation of this species with the same feces resulted in disease in almost all the inoculated animals. In general, subpassage of the MS-1, Phoenix antigen, or CR326 strains of HAV in marmosets (*S. mystax* or *S. rufiventer*) results in increased efficiency of transmission and shortened incubation period. For example, intravenous inoculation of 10 S *mystax* with human acute-phase serum (MS-1 serum) caused elevations in SICD activity in 5 after incubation periods of 49 to 63 days. Passage of the MS-1 strain of virus in this species of marmoset resulted in transmission of disease to all animals in the fifth or higher passage. The incubation period was also shortened to 4 days in animals receiving eighth (or higher) passage of HAV. The shortened incubation period seen in marmosets after passage of MS-1 HAV may reflect either an increase in virus titer in subsequent passages, or an adaptation of the virus to the marmoset. In fact, titration by RIA or HAV in marmoset livers obtained during necropsy has shown that higher passage livers may contain higher titers of HAV.

Chimpanzee Model

As previously mentioned, chimpanzees are highly susceptible to infection with human HAV and develop enzymatic, histopathologic, and serologic evidence of acute hepatitis A. Colony-born chimpanzees have been shown to be uniformly susceptible to infection, whereas wild-caught animals are sometimes immune, that is, positive for anti-HAV. Chimpanzees rarely exhibit frank clinical symptoms during the acute phase of disease, except for lethargy, loss of appetite, occasional diarrhea, and fever. The course of disease in most animals is unremarkable, and none die as a direct result of acute viral hepatitis. There is no evidence of chronic or persistent hepatitis in chimpanzees experimentally infected with HAV.

The incubation period for hepatitis A in chimpanzees experimentally infected by the oral or intravenous route has been shown to range between 9 and 60 days. For example, intravenous inoculation of two chimpanzees with an acute-phase HAV-positive human fecal preparation induced elevated ALT activity 13 and 17 days, respectively, after inoculation. Intragastric inoculation of two other chimpanzees with identical doses of HAV resulted in elevated ALT activity 45 and 53 days, respectively, after inoculation. Although all four animals exhibited elevated ALT activity and seroconverted to HAV, only the two intravenously inoculated animals had demonstrable HAV in feces and relatively high concentrations of virus in liver during the acute phase of disease. From these studies, it appears that the oral route of inoculation is far less efficient than the intravenous route in establishing hepatitis A infection in chimpanzees.

Examination of acute-phase liver biopsies from chimpanzees experimentally infected with HAV reveals histologic alterations similar to those seen in human hepatitis. Generally, the histologic changes observed in acute-phase chimpanzee liver biopsies are more mild than those seen in humans and are most evident near the peak of ALT activity. Numerous studies have also shown that these hepatic lesions occur maximally 1 to 2 weeks after the peak excretion of HAV in feces and the appearance of IgM anti-HAV in serum. One study showed that early histopathologic changes in chimpanzee liver consisted of accumulations of lymphocytes and macrophages in the lobular periphery with replacement of hepatocytes. Concurrent portal tract enlargement with dense infiltrates of mononuclear cells (with lymphocytes predominant over histiocytes) was also observed in the early stages of disease. Hepatocellular degeneration with focal necrosis followed; variations in hepatocyte size and staining characteristics occurred, with subsequent progression to the formation of acidophilic bodies. Portal inflammation was maximal at or just after the peak of ALT activity, and was associated with the proliferation of bile ductules. In some places the portal inflammatory exudate extended into the parenchyma and was associated with erosion of the limiting plate. Hepatic vein tributary lesions were not observed, even at the height of histopathologic changes. Septal or intralobular fibrosis was absent.

SEROLOGIC TESTS FOR HAV ANTIGEN AND ANTI-HAV

Hepatitis A Virus and Anti-HAV

HAV antigen and anti-HAV can be detected by a variety of established methods, including IEM, complement fixation (CF), immune adherence hemagglutination (IAHA), RIA, enzyme immunoassay (EIA), immunofluorescence (IF), and immunoperoxidase staining. Although IEM is a relatively sensitive and specific procedure for the detection of virus particles and antibody, it is a time-consuming method and must be performed by an experienced electron microscopist. The development of simpler serologic tests for HAV and anti-HAV, including CF, IAHA, RIA, and EIA, has made it feasible to routinely screen laboratory specimens. RIA and EIA procedures, in particular, are suitable for the analysis of large numbers of specimens and may be readily performed by a laboratory technician. Immunofluorescent and immunoperoxidase procedures have a more limited application and are generally reserved for research purposes. The detection of HAV by IF in liver biopsies from experimentally infected marmosets and chimpanzees has helped reveal the temporal relationship of HAV antigen appearance in hepatocytes to the development of elevated enzymes, shedding of virus in feces, and occurrence of IgM anti-HAV. IgG anti-HAV conjugated to horseradish peroxidase has been used as an immunologic probe to reveal the presence of free and intravesicular particles in hepatocyte cytoplasm that are immunologically identical to HAV.

Description of Serologic Tests

The early application of IEM, IAHA, and RIA procedures to the diagnosis of hepatitis A virus infection clearly showed that not all acutely infected patients had detectable virus in stool, nor could useful levels of HAV antigen be detected in serum. The variable excretion patterns of HAV in stools from infected patients and the inability of even the most sensitive serologic tests to detect HAV antigen in all acute-phase stools made it necessary to resort to measurement of an individual's serologic response. Seroconversion to HAV infection was first documented by an IEM technique in which serum anti-HAV was crudely quantitated by noting the amount of antibody that bound to reagent HAV particles. This early procedure, however, was not practically useful in the serodiagnosis of acute hepatitis A, since paired acute and convalescent sera were required for the determination of an antibody response. Unfortunately, CF, IAHA, RIA, and ELISA procedures also suffer from the marked disadvantage of requiring paired sera for the diagnosis of hepatitis A.

As with most other virus infections, infection with HAV induces a virus-specific IgM antibody response during the acute phase of disease. This antibody response occurs in all infected individuals, although the duration of anti-HAV IgM and the relative proportions of acute-phase IgM- and IgG-specific anti-HAV may vary from person to person. It is evident from Figure 6 that any serologic test that can discriminate between IgM (acute-phase) and IgG (convalescent-phase) anti-HAV antibody is suitable for the diagnosis of acute viral hepatitis A. A variety of RIA and enzyme-linked immunosorbent assay (ELISA) procedures have been adapted for the purpose of differentially and specifically detecting anti-HAV IgG and/or IgM. Brief descriptions of the basic principles of these tests follow:

Solid-Phase (Antibody Sandwich) RIA for HAV and Anti-HAV

HAV antigen or antibody to HAV can be detected by solid-phase RIA procedures, as shown in Figure 8. IgG containing anti-HAV activity is used to coat the wells of a polyvinyl microtiter U-plate (Fig. 8A) when HAV antigen is to be determined in a specimen. After the wells of the U-plate have been coated with antibody, aliquots of the test specimens (normally stools) are loaded into the wells. HAV antigen present in stool combines with specific anti-HAV; after incubation, the U-plate is washed to remove unreacted materials. In the final step, aliquots of ^{125}I-labeled IgG are added to the wells; HAV antigen present in any of the wells will react with the radiolabeled "probe." After the second incubation period, the wells are again washed to remove unreacted probe, cut out with scissors, and placed in a γ

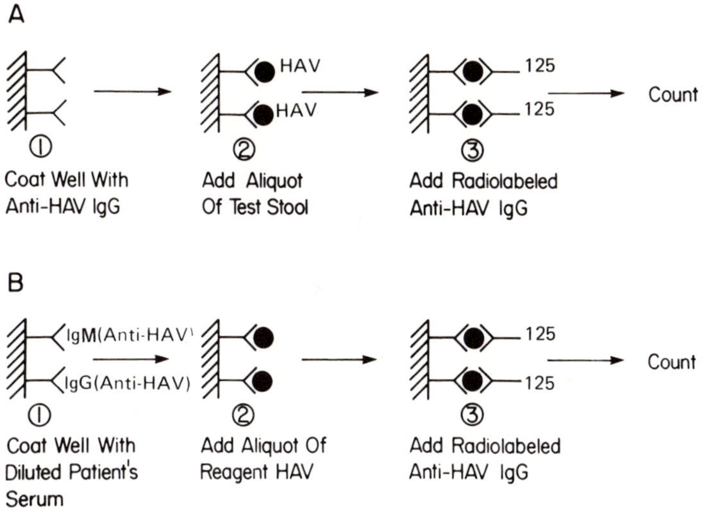

Figure 8 Solid-phase radioimmunoassay for HAV antigen (A) and anti-HAV antibody (B). (From D. W. Bradley, *Journal of Virological Methods* 2:37, 1980.)

spectrometer for counting. The positivity of a given test specimen is determined by computing the ratio of the test specimen's cpm (P) to that of the average cpm of five negative stools (N). Statistical analysis of large numbers of P/N values has shown that a ratio of 2.1 adequately discriminates between negative ($P/N < 2.1$) and HAV-positive ($P/N \geq 2.1$) stools. The RIA for anti-HAV is performed in basically the same manner (Fig. 8B), except that the patient's serum is used to coat the U-plate well. Test sera are normally diluted 1:1000 in phosphate-buffered saline (PBS) and then aliquoted into U-plate wells. A standard HAV antigen (either a stool suspension or a liver homogenate) is used in place of the stool specimen described above. The rest of the test procedure is the same as that of the RIA for HAV antigen. Sera that contain anti-HAV activity (either IgM- or IgG-specific) will react with the reagent HAV antigen; radiolabeled probe will subsequently combine with the HAV antigen retained in wells coated with anti-HAV-positive sera. The positivity of a serum is determined by computing a P/N ratio, where N is the average cpm of five negative sera. Sera are considered to be positive for anti-HAV if $P/N \geq 2.1$.

Enzyme-Linked Immunosorbent Assay for HAV and Anti-HAV

In place of the radiolabeled IgG probe used in RIA procedures, ELISA techniques employ IgG conjugated to an enzyme, such as alkaline phosphatase or horseradish peroxidase (HRP). The enzyme-conjugated IgG serves as indicator in the assay by producing a colored reaction product from a suitable substrate. One of the most commonly used indicator enzymes is HRP; substrates for this enzyme include 5-aminosalicylic acid (5'-AS), ortho-phenylenediamine (OPD), and 2,2'-azinodi(3-ethylbenzthiazoline-6-sulfonate) (ABTS). 5'AS and OPD both yield yellow reaction products, while ABTS gives a blue-green product. As can be seen in Figure 9, the ELISA procedure, shown here in the configuration of an antibody sandwich, is basically the same as that of the solid-phase RIA procedure (described above). Although the sensitivity of the ELISA procedure may be somewhat less than a comparably configured RIA, ELISA procedures in general have several distinct advantages over their RIA counterparts. These advantages include long shelf life of kit components (conjugate, substrate, and so on), as well as the economy and simplicity of the test itself, since a relatively inexpensive spectrophotometer can be used to measure the colored reaction product.

Alternative configurations of the basic RIA and ELISA procedures discussed above have been described, including a commercially produced RIA kit* for the detection of anti-HAV. As of October 1980, this kit is the only one that is commercially available. However, other manufacturers are soon expected to market RIA or ELISA kits for the detection of anti-HAV. The following discussion of the determination of IgM- and IgG-specific anti-HAV by RIA and ELISA procedures includes a description of a modified HĀVAB procedure that can be used to differentially detect anti-HAV IgM and IgG.

μ-Chain (IgM) Blocking RIA

A modification of the solid-phase RIA procedure for the detection of anti-HAV was shown to be useful in the differentiation of acute- and convalescent-phase hepatitis A sera. The modified RIA employed an additional step in which goat anti-human IgM (μ-chain specific) was used to block binding of reagent HAV to test serum IgM coated on the U-plate well. Acute-phase sera containing primarily anti-HAV IgM were shown to be blocked by anti-IgM antibody, yielding reduced cpm compared to the unblocked control. Convalescent sera containing

---

*HĀVAB (Abbott Laboratories). All commercial names are used for identification only and their mention does not constitute endorsement by the U.S. Department of Health and Human Services.

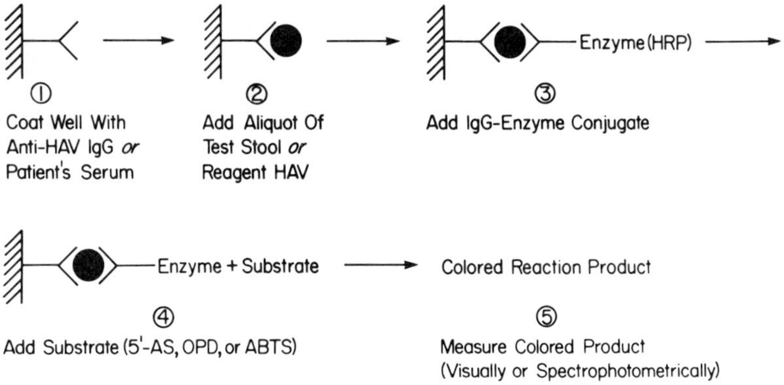

Figure 9 Enzyme-linked immunosorbent assay (ELISA or EIA) for HAV and anti-HAV. (From D. W. Bradley, *Journal of Virological Methods* *2*:38, 1980.)

primarily, if not entirely, anti-HAV IgG were not blocked by the addition of anti-IgM antibody to the U-plate well. Although this μ-chain-blocking RIA procedure was found to be useful in the serodiagnosis of acute viral hepatitis A, it required the analysis of two or or more dilutions of the test serum and use of a highly specific blocking antibody. (It should be noted that many commercial preparations of goat, rabbit, or horse anti-human IgM were also found to contain anti-human IgG activity.)

Modified Competitive Binding RIA for IgM Anti-HAV

Abbott Laboratories introduced a radioimmunoassay kit (HĀVAB) for the determination of anti-HAV in serum in 1979. The standard HĀVAB procedure, however, does not differentially detect anti-HAV IgM and IgG and cannot be used to serodiagnose acute viral hepatitis A using a single serum specimen. The principle of the standard HĀVAB procedure is shown in Figure 10. The HĀVAB test is a solid-phase competitive-binding RIA based on the use of a polystyrene bead coated with HAV derived from infected primate (marmoset) liver. The solid-phase HAV can combine with anti-HAV IgM or IgG in a test serum as well as with the ^{125}I-labeled IgG anti-HAV used as the test probe. If a patient's serum is negative for anti-HAV, addition of an HAV-coated bead to a mixture of this serum and [^{125}I]IgG anti-HAV will result in maximum binding of the [^{125}I]IgG probe. If a patient's serum is positive for anti-HAV, however, addition of an HAV-coated bead to a mixture of test serum and [^{125}I]IgG anti-HAV will result in a decreased

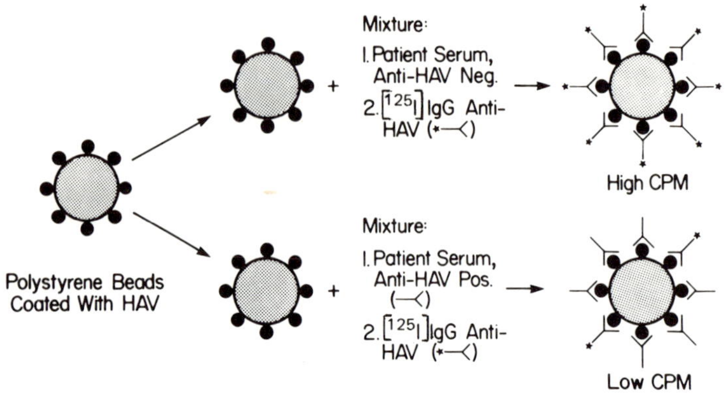

Figure 10 Commercially available radioimmunoassay for the detection of anti-HAV antibody. (From D. W. Bradley, *Journal of Virological Methods* 2:40, 1980.)

binding of the [^{125}I]IgG probe. The Patient's anti-HAV, whether it is of the IgM or IgG class of antibody, will compete with [^{125}I]IgG for HAV binding sites. Presumably, higher titered sera will result in greater competition with [^{125}I]IgG for HAV binding sites.

A modified form of the standard HĀVAB test was developed for differentiation of IgM- and IgG-specific anti-HAV in a single acute-phase serum specimen. The modified HĀVAB procedure uses protein A from *Staphylococcus aureus* cells to preferentially absorb anti-HAV IgG from a patient's serum. It is well-known that protein A has a has a much higher affinity for IgG than for IgM and that it will react with approximately 95% of the IgG species (subclasses 1, 2, and 4) present in serum. Under appropriate test conditions, the predominant anti-HAV antibody type in a patient's serum can be readily determined by the modified HĀVAB procedure (referred to as HĀVAM). Invalid test results (i.e., false positive anti-HAV IgM) may result if improper dilutions or aliquots of serum are used. Protein A used for the absorption of serum must also be titered to determine its potency, since different lots and *S. aureus* strains vary in their capacity to absorb IgG. One of the advantages of the HĀVAM procedure is its ability to determine the predominant anti-HAV antibody type, whether it is IgM or IgG, in contrast to μ-chain-specific anti-HAV procedures that yield positive test results only if anti-HAV IgM is present (to be discussed in detail below). Another potential advantage of the HĀVAM test is its relatively moderate sensitivity for anti-HAV IgM. Sera collected from infected individuals more than 4 weeks after the onset of symptoms are often negative for anti-HAV IgM, and almost all sera collected 2 to 3 months after onset of illness are negative for anti-HAV IgM. This relatively narrow anti-HAV IgM "window" facilitates diagnosis of a patient's illness, since there is a clear temporal relationship of illness to positivity for anti-HAV IgM. The HAVAM procedure has been successfully used to define major epidemics of viral hepatitis A in Burma, France, and the United States, and has proved to be an accurate and reliable laboratory test when properly performed.

μ-Chain-Specific RIA and ELISA Procedures for the Detection of IgM Anti-HAV

Anti-HAV IgM can be specifically detected by RIA and ELISA procedures that incorporate the use of a μ-chain-specific antibody that will bind human IgM. Several reports, in fact, have described μ-chain-specific anti-HAV tests; the basic principles of these tests are briefly reviewed below:

Solid-Phase RIA for Anti-HAV IgM The principles of these tests are shown in Figure 11. Rabbit or goat antiserum to human IgM (μ-chain specific) is used to coat the wells of a polyvinyl microvinyl microtiter plate. Diluted patient serum is then added to one of the U-plate wells. IgM antibody present in the patient's serum will bind to the μ-chain-specific antibody coated on the well, forming a double-antibody sandwich. A specified quantity of reagent HAV is added to the well; anti-HAV IgM bound to the solid-phase anti-μ antibody will then bind HAV. Finally, radiolabeled IgG containing anti-HAV activity is added

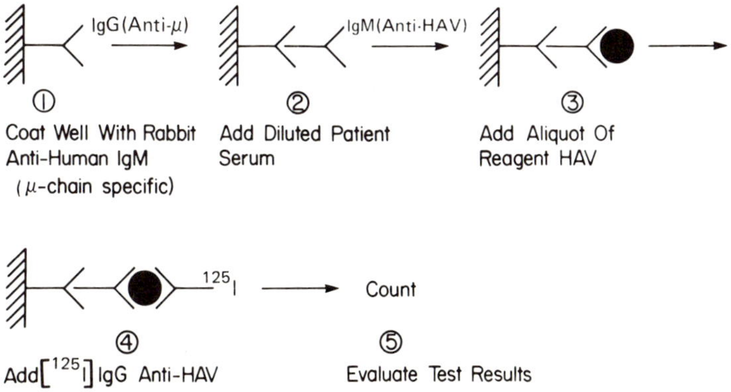

Figure 11 Solid-phase RIA for IgM (μ-chain-specific) anti-HAV. (From D. W. Bradley, *Journal of Virological Methods* 2:41, 1980.)

to the well, where it will bind to HAV. Test sera containing anti-HAV IgM will yield high cpm, whereas sera negative for anti-HAV IgM will yield low cpm. Positivity of a serum for acute-phase (IgM) anti-HAV can be evaluated by using suitable negative and positive control sera. Abbott Laboratories has recently introduced a μ-chain-specific RIA for the detection of IgM anti-HAV, called HĀVAB-M, that embodies the above test principles.

The solid-phase RIA for anti-HAV IgM has been shown to be specific and highly sensitive. In fact, in one study of 60 sera collected from different patients at various times after the onset of icterus, 25 (42%) were found to be positive for anti-HAV IgM by an experimental RIA procedure as long as 6 months after onset, while 13 (22%) were judged to be positive for anti-HAV IgM 52 weeks after onset. These findings demonstrate that the sensitivity of the RIA test for IgM anti-HAV is greater than that of the modified HĀVAB procedure. A potential disadvantage of this test procedure, however, would appear to be related to its extreme sensitivity, since a patient with acute non-A hepatitis could conceivably be positive for anti-HAV IgM a year or more after infection with HAV. In this regard, it would be necessary to couple the test results with the patient's history of viral hepatitis to resolve the question of acute hepatitis A virus infection. Alternatively, test conditions could be altered to reduce the inherent sensitivity of the RIA procedure (by dilution of the test serum) and eliminate the possibility of a "false positive" result and subsequent misdiagnosis. It should also be noted that rheumatoid factor, a potentially interfering antibody, was not found to interfere in the HĀVAB-M test for anti-HAV IgM.

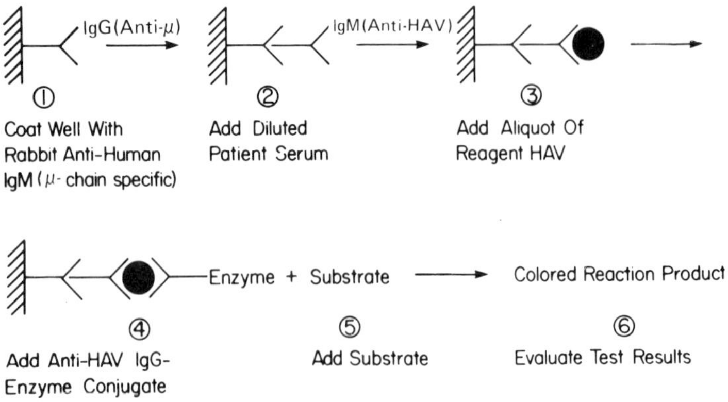

Figure 12 ELISA for IgM (μ-chain-specific) anti-HAV. (From D. W. Bradley, *Journal of Virological Methods* 2:42, 1980.)

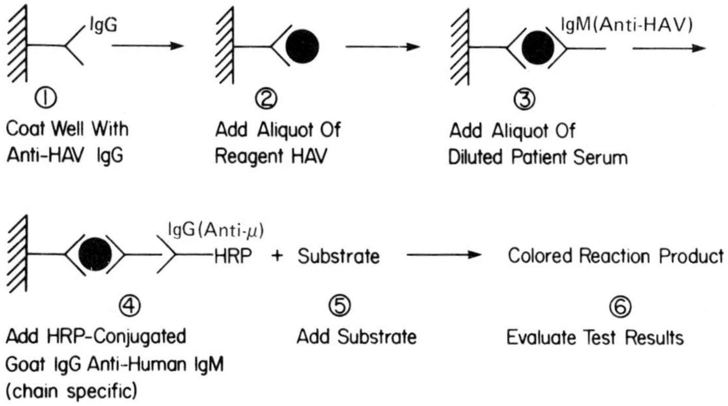

Figure 13 ELISA for IgM (μ-chain-specific) anti-HAV. (From D. W. Bradley, *Journal of Virological Methods* 2:43, 1980.)

ELISA for Anti-HAV IgM: Type I The principles of this test are shown in Figure 12 and are basically the same as those described for the RIA procedure, except that HRP-conjugated IgG containing anti-HAV activity is used as the test probe. Test sera positive for anti-HAV IgM will yield a yellow-colored reaction product in the ELISA procedure, while sera negative for anti-HAV IgM will not yield a colored reaction product. Rheumatoid factor was not found to give a positive reaction in the ELISA test for anti-HAV IgM. It is presently unclear whether the sensitivity of the ELISA procedure for anti-HAV IgM will be problematic in the interpretation of some test results. As previously mentioned, however, intentional reduction in the sensitivity of this test procedure could be used to circumvent such difficulties.

ELISA for Anti-HAV IgM: Type II A variation of the above ELISA procedure is shown in Figure 13. The wells of a microtiter plate are coated with the standard IgG containing anti-HAV activity. Reagent HAV is then added to each well, where it will bind to anti-HAV. Aliquots of diluted patient sera are added to the wells; anti-HAV IgM or IgG present in the sera will bind to the solid-phase HAV. Anti-HAV IgM bound to HAV is detected by an enzyme conjugate consisting of HRP linked to a goat anti-human IgM-specific (μ-chain) IgG. Sera positive for anti-HAV IgM will yield a yellow-colored reaction product that can be quantitated visually or spectrophotometrically.

CONTROL AND PREVENTION

The control and prevention of hepatitis A, from the epidemiologic viewpoint, has centered upon the identification of the infective patient or other source of contamination, interdiction of spread by removal of the contaminating source, and use of immunoglobulins (IG) for the prevention of illness in susceptible contacts. Prompt notification of cases to health authorities and early epidemiologic investigation of common-source outbreaks with identification of index cases, food vehicles, and susceptible individuals, is a key feature to effective control and prevention of disease. It is estimated that less than 10% of hepatitis cases in the United States are reported to health authorities, even where laws mandating such reporting exist. Clinicians should be aware of the critical role they must play in the surveillance system established to control this disease. Since hepatitis A is not a significant nosocomial infection, revised guidelines for the environmental management of hospitalized patients with this disease have been formulated which emphasize careful attention to precautions generally used in the handling of urine and feces of all hospitalized patients.

The only method to achieve widespread effective long-term prevention of hepatitis A is active immunization. Fortunately, the current success in cultivation of HAV in cell culture systems offers the likelihood of vaccine development within the next several years. Until such development, however, use of passive prophylaxis with IG will constitute a mainstay of hepatitis A control.

Numerous studies have verified that standard IG, also called immune serum globulin (ISG), may confer protection when administered before or soon after exposure to HAV. Effectiveness of standard IG is based upon its specific content of anti-HAV, although the minimum titer of anti-HAV necessary to protect has not been determined. Practically all currently manufactured standard IG contains some anti-HAV and may be expected to confer protection. When administered within 14 days of exposure, standard IG may prevent illness in up to 80% of exposed individuals. Also, because this globulin presumably does not suppress inapparent infection, primary antibody response (passive-active immunity) produced under cover of passively acquired antibody may be expected to result in acquisition of long-term immunity.

The dosage patterns of standard IG in common use may vary with exposure setting. In the postexposure setting, a single dose of 0.02 ml/kg body weight is generally recommended. In the preexposure setting, the dosage varies with body weight and length of time protection is required. Specific recommendations and doses of standard IG for hepatitis A prophylaxis have been published in the United States by the Public Health Service Advisory Committee on Immunization Practices and are summarized as follows:

Postexposure Prophylaxis

Close Personal Contact Close personal contact, as among permanent or temporary household residents, is important in the spread of hepatitis A. Standard IG is recommended for all household contacts who have not had hepatitis A.

School Contacts Routine IG administration to casual pupil or teacher contacts of a case in school is not indicated. However, when epidemiologic investigation has shown that a school- or classroom-centered outbreak exists, standard IG may be administered to persons at risk.

Day Care Centers Day care centers having diapered attendees constitute important sources of HAV infection. Once hepatitis A has been documented, standard IG should be administered to all staff and attendees.

Institutional Contacts Conditions in institutions, such as prisons and facilities for the mentally retarded, favor the spread of hepatitis A. Administration of standard IG to residents and staff contacts of hepatitis A cases is recommended.

Common-Source Exposure Standard IG should be effective in preventing food- or water-borne disease. However, exposure is often recognized too late for IG to be effective. In some instances, potential food contamination is suspected when an HAV-infected foodhandler is diagnosed, but before the occurrence of disease in exposed individuals. Depending on whether IG administration could be accomplished within 2 weeks of exposure and whether the foodhandler was involved in the handling of uncooked foods or foods after cooking, IG may be recommended for exposed individuals. IG is recommended for fellow foodhandlers who work and consume food in the infected kitchen.

Preexposure Prophylaxis

In susceptible individuals traveling as tourists to endemic areas of the world where sanitary level is poor, particularly those who bypass ordinary tourist routes, hepatitis A may constitute a risk. Avoidance of potentially contaminated water and uncooked foods is the primary preventive measure here. However, standard IG may be considered prior to departure at a single dose of 0.02 ml/kg of body weight. Individuals planning to stay for extended periods of more than 3 months or who plan to reside in tropical areas or developing countries where hepatitis A is common are at greater and continuing risk of acquiring hepatitis A. An injection of standard IG in a dose of 0.05 ml/kg body weight is recommended for them prior to departure and again at 4 to 6 month intervals for the duration of their stay.

SUGGESTED READINGS

Bradley, D. W., H. A. Fields, K. A. McCaustland, E. H. Cook, C. R. Gravelle, and J. E. Maynard. Biochemical and biophysical characterization of light and heavy density hepatitis A virus particles: Evidence HAV is an RNA virus. *J. Med. Virol. 2*: 175-187, 1978.

Bradley, D. W., F. B. Hollinger, C. L. Hornbeck, and J. E. Maynard. Isolation and characterization of hepatitis A virus. *Am. J. Clin. Pathol. 65*:876-889, 1976.

Bradley, D. W., J. E. Maynard, S. H. Hindman, C. L. Hornbeck, H. A. Fields, K. A. McCaustland, and E. H. Cook. Serodiagnosis of viral hepatitis A: Detection of acute-phase immunoglobulin M anti-hepatitis A virus by radioimmunoassay. *J. Clin. Microbiol.* 5:521-530, 1977.

Bradley, D. W., and J. E. Maynard. Serodiagnosis of viral hepatitis A by radioimmunoassay. *Lab. Manag. 16*:29-34, 1978.

Dienstag, J. L. The pathobiology of hepatitis A virus. *Int. Rev. Exp. Pathol. 20*:1-48, 1979.

Dienstag, J. L., H. Popper, and R. H. Purcell. The pathology of viral hepatitis types A and B in chimpanzees. A comparison. *Am. J. Pathol. 85*:131-148, 1976.

chapter 5

Non A–Non B Viral Hepatitis

DANIEL W. BRADLEY and JAMES E. MAYNARD

U.S. Public Health Service
Centers for Disease Control
Phoenix, Arizona

EPIDEMIOLOGY AND ETIOLOGY

Rapid advances in the field of viral hepatitis during the last two decades, particularly the development of highly sensitive and specific laboratory techniques for the diagnosis of infections caused by hepatitis A and hepatitis B viruses, have facilitated the recognition of a third category of hepatic infections not due to either of these agents. The finding that the vast majority of the infections are also not due to other viral agents which may cause hepatic injury, such as cytomegalovirus or Epstein-Barr virus, has resulted in their designation as non A-non B viral hepatitis. Because no serologic techniques have been developed which can consistently detect markers of infection, the diagnosis is currently one of exclusion, to be made only after infection with the other known viral agents causing hepatitis have been ruled out. That the disease is, in fact, caused by transmissible agents of presumptive viral nature, has been amply proven in experimental transmission studies in nonhuman primates.

Despite the lack of laboratory techniques for specific diagnosis of non A-non B hepatitis or for the study of seroepidemiologic patterns of infection, investigations of this disease in a variety of areas of the world, based on case designation by exclusion of other viral etiologies, have resulted in the collection of significant information regarding the epidemiology of this infection. Furthermore, epidemiologic observations have contributed to current hypotheses regarding the number of possible agents responsible for non A-non B hepatitis.

In North America and Europe, non A-non B viral hepatitis first appeared within the context of continuing posttransfusion hepatitis despite the institution of rigorous testing procedures to eliminate the presence of hepatitis B virus in donor blood. In the United States the incidence of posttransfusion hepatitis in multiply transfused patients has been currently estimated at 8 to 17%, of which approximately 90% of cases cannot be attributed to known hepatotrophic viruses. As early as 1962, it was reported that approximately 25% of posttransfusion hepatitis had incubation periods too short for hepatitis B. In 1971 and 1974, two U.S. studies of transfusion-associated hepatitis, which failed to detect serologic markers of hepatitis B virus (HBV) in many patients, also ruled out an association between cytomegalovirus or Epstein-Barr virus infection, and suggested that other agents might play an etiologic role in this disease. A further study also ruled out hepatitis A virus as a cause of non B posttransfusion hepatitis.

The fact that donor blood is capable of transmitting non A-non B hepatitis indicates that a chronic blood carrier state for the responsible agent (or agents) must, indeed, exist in humans and constitute an important reservoir for transmission of infection. In addition to blood itself, other blood products or plasma derivatives, such as factor VIII and factor IX complexes, as well as platelets, have been shown to cause disease. In a retrospective study of blood donors implicated in the transmission of non A-non B hepatitis to recipients, none had a prior history of clinically recognizable hepatitis, 93% had never received a blood transfusion, and 80% had normal levels of serum aminotransferase. The epidemiologic implications of these findings, as shown in Table 1, are (1) that a chronic asymptomatic carrier state for non A-non B hepatitis exists in humans, (2) that blood and blood or plasma products are reservoirs for transmission of infection, (3) that a significant frequency of acute asymptomatic infection occurs, (4) that non A-non B agents may remain in the circulation for significant periods after recovery from clinical illness, and (4) that transmission of infection by means other than blood transfusion is a frequent occurrence.

With regard to etiology, epidemiologic evidence for the existence of at least two separate non A-non B agents derives from observation of disease with differing incubation periods. Disease of short incubation (4 to 40 days) was first reported in 1962 following transfusion of blood, although hepatitis B could not be ruled out due to lack of diagnostic test procedures. However, in 1977, an epidemic of non A-non B disease due to platelet infusions with short incubation (14 to 28 days) was reported. This outbreak had originally been misdiagnosed as hepatitis A because of the short incubation period. In addition, such short incubation period disease has been reported on two separate occasions following infusions of factor VIII concentrates. In contrast, long incubation period non A-non B hepatitis (3 to 4 months) has been

Table 1 Epidemiologic Inferences from Studies of Blood Donor Transmitted Non A-Non B Hepatitis

1. Existence of a chronic asymptomatic carrier state in humans.
2. Human blood and blood products are reservoirs for transmission of infection.
3. Frequent occurrence of asymptomatic infection.
4. Continuing infectivity after recovery from acute illness.
5. Frequent transmission by methods other than blood transfusion.

observed following transfusion both of blood and factor IX concentrates. Since dose-response dynamics play a role in the determination of incubation period, it is difficult to fully evaluate the range of such periods attributable to differing agents in the absence of specific laboratory tests which can quantitate inoculum size. It is interesting to note, however, that one experimental cross-challenge study has been done in champanzees in which animals convalescent from a short incubation period factor VIII-induced infection subsequently again developed non A-non B hepatitis with a long incubation period when inoculated with factor IX materials which had previously been shown to produce long incubation period disease. Conversely, a chimpanzee convalescent from long incubation period factor IX-induced non A-non B hepatitis again developed disease with a short incubation period when rechallenged with the above factor VIII materials. This information, together with clinical data which document the occurrence of multiple bouts of discreet hepatitis in the same individual for whom at least two distinct episodes could not be accounted for by known hepatotrophic viruses, supports the epidemiologic inferences on the multiple etiologies of non A-non B hepatitis. It now seems irrefutable that at least two distinct agents are responsible for the production of this disease.

The geographical distribution and impact on total viral hepatitis incidence of non A-non B agents has been the subject of considerable interest on a worldwide basis. In 1977, a study of cases of acute viral hepatitis hospitalized in Los Angeles revealed that one-half of the non B hepatitis cases could not be attributed to hepatitis A. Since that time, sophisticated diagnostic techniques to rule out infection with other known hepatitis viruses have been applied to hospitalized cases of disease in a number of countries. Table 2 summarizes largely unpublished data from several of these countries. With the exception of Germany, in which a series of both adults and children have been studied, most results derive from adult hospital admissions and indicate

Table 2 Distribution of Hospitalized Cases of Acute Viral Hepatitis by Country and Etiology

		Etiology		
Country	N	A (%)	B(%)	Non A-non B (%)
Australia	1006	29	50	21
England	243	31	51	18
Germany				
Adults	155	15	55	30
Children	168	80	14	6
Greece	216	10	76	14
Japan	102	20	32	48
Malawi	33	15	64	21
Sweden	473	23	63	14
Switzerland	115	?	?	21
USA	140	18	56	26

that non A-non B disease may account for between 14 and 48% of such adult cases. In Germany, Greece, Japan, Africa, and the United States, non A-non B viral hepatitis appears to be more important than hepatitis A in its contribution to overall adult disease, and in Japan may even exceed hepatitis B in relative frequency. The German data substantiate the common belief that hepatitis A is the most important contribution to the etiology of childhood illness, but also indicates that children are not exempt from the acquisition of non A-non B illness. In general, non A-non B hepatitis has been identified wherever it has been sought and it is now clear that its distribution is worldwide. Furthermore, it appears to contribute significantly to the occurrence of acute adult hepatitis and may cause disease in children.

Since most of the above data derive from hospitalized cases and do not, therefore, represent all community-acquired acute viral hepatitis, the Centers for Disease Control in the United States has recently initiated an etiologic and epidemiologic surveillance of acute viral hepatitis in five community health constituencies (counties) chosen because of the excellence of their disease surveillance infrastructures. Acute cases reported to these local health authorities have been subjected to serologic diagnosis and epidemiologic investigation. The criteria for serologic diagnosis require use of most or all of the currently available sensitive test methodologies for the diagnosis of hepatitis A and B. These methodologies are such, however, that the diagnosis of hepatitis A or B can be excluded using a single serum specimen collected within 6 weeks of the onset of illness.

Results of preliminary analysis of data from 298 cases of disease accessed into the study between November 1979 and March 1980 showed that 31% of the cases were hepatitis A, 38% hepatitis B, and 31% non A-non B hepatitis. The bulk of reported disease occurred in the age groups 15 to 34 years, regardless of etiologic classification. Although jaundice predominated in all groups, undoubtably due to its being the principal feature upon which clinical diagnosis is made in these community settings, well over half of all cases, regardless of etiologic category, did not end in hospitalization. Thus, the sample appeared to be more representative of disease occurring in the community than would be the case for hospitalized patients alone. While prior history of contact with another hepatitis case, as well as foreign travel, mostly in areas where poor sanitation prevails, were associated with hepatitis A, both hepatitis B and non A-non B hepatitis were, in contrast, more frequently associated with history of illicit drug use, prior surgery, transfusion, and hospitalization. While hepatitis B was associated with a history of homosexual activity, non A-non B hepatitis did not appear to be similarly related. With this exception, the epidemiologic characteristics of non A-non B hepatitis generally appeared to be more similar to those of hepatitis B than hepatitis A.

Of importance is the finding that a substantial number of non A-non B cases could not be related to known percutaneous exposure to blood in the 6 months prior to onset of illness. In fact, in about 50% of cases, no clear history of any evident contact could be elicited. This suggests that as yet not fully defined nonparenteral routes are of major significance in the transmission of non A-non B hepatitis. Close contact within families and between sexual partners have been reported to be associated with transmission of this disease. It is probable that more obscure but still potent mucosal or skin contact with presumably infectious bodily secretions, such as urine, saliva, or semen, may, as with hepatitis B, be important mechanisms of transmission.

Although epidemiologic data from Europe and North America do not suggest that the fecal-oral route is a significant mode of transmission for non A-non B hepatitis in these areas, recent data from India suggest that at least one form of such disease may be water borne. The largest known published outbreak of viral hepatitis occurred between December 1955 and January 1956 in New Delhi, India. During this period, an estimated 29,000 icteric cases occurred subsequent to a large-scale fecal contamination of the city water supply system. The compressed and unimodal nature of the curve of epidemic illness was also consistent with a common-source contamination. This epidemic has long been attributed to hepatitis A virus despite certain acknowledged unusual epidemiologic features. These included a somewhat longer than usual incubation period (mean of about 40 days), a high attack rate in adults (2.9% in the age group 15 to 39 years of age as compared with a rate of 1.2% for the group aged 14 years and under) and an unusually high mortality rate in pregnant women (case fatality estimated at 10%). Both acute illness and convalescent-phase sera collected from individuals who acquired hepatitis during this outbreak and stored under optimal conditions have recently been examined for evidence of infection by known hepatitis viruses. Results from these investigations now indicate that neither hepatitis A nor B viruses could have accounted for the epidemic, and it is likely that an as yet undefined non A-non B viral agent was the cause of the outbreak. More recently, an epidemic of viral hepatitis ascribed to contaminated water and occurring in late 1978 in several villages in the Kashmir Valley of India has been reported. The epidemic exhibited a short incubation period, high attack rates in adults, a severe clinical course in pregnant women, and a high frequency of fulminant hepatitis. Extensive examination of serum samples taken from individuals in the acute phase of illness showed no evidence of concurrent infection by either hepatitis A or B viruses. Again, the possibility that a similar non A-non B agent transmissible by the fecal-oral route was responsible for this outbreak must be seriously entertained.

The occurrence of explosive viral hepatitis in India, spread by person-to-person contact as well as common vehicle, with a high attack rate in adults despite high levels of background immunity to hepatitis A, and with severe clinical manifestations in pregnant women, has long been attributed to hepatitis A virus. Explanations for the observed variances in typical epidemiologic patterns have been ascribed to either existence of differing strains of hepatitis A virus with variations in antigenic composition and virulence, or to reinfection predicated on high dose of inoculum, waning immunity, and impaired host response. The probable existence of a non A-non B agent responsible for these outbreaks offers a more tenable explanation for their occurrence, since different antigenic variants of hepatitis A virus are not known to exist, and there is ample evidence for the occurrence of lifelong immunity following a single hepatitis A virus infection. Since the epidemiologic pattern for these outbreaks is clearly different from that for non A-non B hepatitis seen in North America and Europe, the possibility exists that a third viral agent, which has not yet penetrated other areas of the world, may be added to the non A-non B etiologic category. The apparent geographical localization of this form of epidemic non A-non B hepatitis to the India subcontinent has yet to be more fully substantiated.

Table 3 General Epidemiologic Characteristics of Non A-Non B Hepatitis

1. *Reservoir*
   Human (nonhuman primates?)
   Chronic carrier state
2. *Etiology*
   Viral (presumed)
   At least two and probably three distinct agents
3. *Incubation period*
   Variable: 5-150 days
   Evidence exists for both a short (5-40 days) and long (40-150 days) incubation illness
4. *Distribution*
   Worldwide: sporadic endemic, (North America, Europe, Mediterranean, Japan, Africa); endemic (India)
5. *Incidence*: accounts for 15 to 50% of adult sporadic endemic hepatitis
6. *Modes of transmission*
   Blood transfusion, blood products
   Other forms of percutaneous inoculation
   Close personal contact including sexual
   Water-borne?

From the above review, the general epidemiologic characteristics of non A-non B hepatitis can be summarized as shown in Table 3. The reservoir for infection is the human, although some nonhuman primates are susceptible to infection. The etiology is evidently viral, although a role for other infectious agents, such as viroids, cannot strictly be excluded. There are at least two distinct agents responsible for sporadic endemic non A-non B hepatitis, and possibly a third which may cause epidemic disease in India. Evidence also exists for distinct forms of illness with both short and long incubation periods. Distribution of infection is worldwide and may account for 15 to 50% of acute adult viral hepatitis cases which occur on a sporadic endemic basis. Epidemics presumably due to a non A-non B agent have also been reported from India. Infection is transmitted by blood transfusion, blood product infusions, other percutaneous inoculation, and close personal and possibly sexual contact; water-borne transmission has also been reported. It is hoped that discovery and complete characterization of the agents responsible for this disease will occur rapidly, with development of appropriate serologic tests to provide confirmation and elaboration of the current epidemiologic data.

TRANSMISSION OF DISEASE TO NONHUMAN PRIMATES

Chimpanzees

Animal transmission studies conducted in 1977 and 1978 demonstrated that wild-caught and colony-born chimpanzees (*Pan troglodytes*) are susceptible to infection by human non A-non B hepatitis agent(s). In one study, five wild-caught chimpanzees were intravenously inoculated with serum or plasma in volumes ranging from 3.0 ml to 75.0 ml obtained from four patients and one commercial donor who had previously developed non A-non B hepatitis. The donor and two of the patients who had developed posttransfusion non A-non B hepatitis demonstrated persistently elevated alanine aminotransferase (ALT) activity; one of these patients had chronic active hepatitis (CAH) and the other had chronic persistent hepatitis (CPH). Plasma used for the inoculation of three chimpanzees was obtained from the two patients with CAH and CPH 35 and 42 weeks, respectively, after transfusion, and from the commercial donor at a time when her ALT activity was elevated. Serum was obtained from the remaining two patients with acute non A-non B hepatitis 4 and 6 weeks, respectively, after transfusion of volunteer donor blood shown to be negative by radioimmunoassay (RIA) for hepatitis B surface antigen (H BsAg).

All five inoculated chimpanzees developed elevated ALT activity between 3 and 17 weeks after inoculation. Well-defined enzyme elevations were observed in three chimpanzees inoculated with either acute-phase serum (two animals) or chronic-phase plasma obtained from the patient with chronic persistent non A-non B hepatitis. Three enzymes, alanine aminotransferase, aspartate aminotransferase (AST), and γ-glutamyl transpeptidase (GGTP), were monitored in serial serum specimens from all five chimpanzees. ALT activity in serum was the most consistent and sensitive indicator of liver dysfunction. The major peaks of ALT activity occurred between 15 and 20 weeks after inoculation, with enzyme values ranging between 62 and 265 mU/ml. None of the chimpanzees had serologic evidence of acute infection or reinfection with hepatitis A virus (HAV), HBV, cytomegalovirus (CMV), or Epstein-Barr virus (EBV), and histopathologic changes in acute-phase liver biopsies from all five animals showed at least some evidence of acute hepatitis ranging in severity from borderline to moderate. Liver biopsies from one chimpanzee inoculated with acute-phase human serum showed focal necrosis, variation in size and staining quality of hepatocyte nuclei, increase in sinusoidal lining (Kupffer) cells, infiltration of portal tracts, and in places, erosion of the limiting plate. In general, the histopathologic lesions observed in the chimpanzee liver biopsies were similar to those seen in type B hepatitis, including panlobular involvement and streaks of focal necrosis that extended into the central zone. By contrast, hepatic lesions seen in type A hepatitis tend to be periportal. The severity of disease in non A-non B-infected chimpanzees was not correlated with the volume of the inoculum, nor was there any specific association of the degree of ALT elevation with the inoculation of acute- or chronic-phase serum or plasma.

In a similar study, four colony-born chimpanzees were intravenously inoculated with sera obtained either from a patient with chronic non A-non B hepatitis or from two former blood donors. Both donors were found to have elevated AST activity $1\frac{1}{2}$ and 5 years, respectively, after being implicated in the transmission of non A-non B hepatitis to two recipients. Elevated ALT activity was observed in all four inoculated chimpanzees between 2 and 4 weeks after inoculation. Serial serum specimens were also assayed for AST and isocitrate dehydrogenase (SICD) activities; however, of the two enzymes, only SICD appeared to correlate well with liver histology. Peak aminotransferase activities occurred between 6 and 15 weeks after inoculation. As in the previously described study, none of these chimpanzees exhibited serologic evidence of acute infection or reinfection with HAV, HBV, CMV, or EBV. Histologic lesions in acute-phase liver biopsies from all four animals were similar to those reported in the above study.

The results of these two studies demonstrated that both wild-caught and colony-born chimpanzees can be infected with human non A-non B hepatitis agent(s). Most important was the finding in each study that chronic-phase serum or plasma obtained from individuals with persistent elevations in aminotransferase activity was infectious in chimpanzees. Both studies experimentally supported the observation that acute non A-non B hepatitis can progress to chronic disease accompanied by persistent viremia.

Reports of the transmission of non A-non B hepatitis to chimpanzees were quickly followed by others that showed similar findings. One study demonstrated that chimpanzees could be infected with serum from a donor with elevated ALT activity, as well as serum from donors with normal ALT activity. Of five chimpanzees, two were inoculated with donor sera with normal ALT activity (one by the intravenous route, the other by cannulation of the inferior mesenteric vein) and two were inoculated with sera obtained from two recipients between 21 and 12 days prior to the onset of non A-non B hepatitis. The latter chimpanzees were inoculated by either the intrahepatic or intramuscular route. The remaining animal was inoculated intramuscularly with serum from a donor with slightly elevated ALT activity. All the inoculated chimpanzees developed significantly elevated ALT activity between 25 and 38 days after inoculation; ALT activity remained elevated for an additional 40 to 70 days. As in the previous studies, there was no serologic evidence of infection with HAV, HBV, CMV, or EBV, and liver histopathology was consistent with a diagnosis of acute viral hepatitis. This study provided unequivocal evidence that (1) an infectious agent may be present in blood collected from recipients during the preacute phase of non A-non B hepatitis, and (2) healthy carriers with normal ALT activity exist within the donor population.

Two other studies showed that non A-non B hepatitis could be transmitted to chimpanzees by intravenous inoculation of commercially prepared clotting factors. Two chimpanzees in one study received either an implicated factor IX concentrate or a factor IX concentrate previously used in recipients without apparent complications. Both animals developed elevated ALT activity approximately 10 weeks after inoculation; however, enzyme elevations were less than 60 mU/ml and were of comparatively short duration. In spite of the relatively mild hepatitis observed in these animals, both showed histopathologic changes in acute-phase liver biopsies indicative of acute hepatitis. Serial liver biopsies from one chimpanzee showed some evidence of progression to chronic liver damage.

A second study showed that non A-non B hepatitis could be transmitted to chimpanzees by intravenous inoculation of factor VIII (antihemophilic) materials implicated in the transmission of non A-non B hepatitis to two geographically isolated patients. The implicated factor VIII materials were commercially prepared from three pools, each containing several thousand units of donor plasma screened by RIA for HBsAg. It should be noted that unlike many other blood products, factor VIII, as well as factor IX, is not treated by heating at 60°C for 10 h to inactivate residual virus (e.g., HBV). Therefore, it is not surprising these materials have a high potential for the transmission of viral agents, including hepatotrophic viruses. Two chimpanzees were each intravenously inoculated with 30 ml of one implicated factor VIII lot (animals 771 and 921), and two others were each intravenously inoculated with 5.0 ml of a pool of two other implicated factor VIII lots (animals 908 and 828). All four animals developed significant elevations in ALT activity after infusion of the factor VIII materials (Fig. 1). The incubation periods to the first obvious enzyme elevations varied between 18 and 53 days. It is interesting to note that two animals inoculated with identical aliquots of the same factor VIII lot (animals 771 and 921) exhibited markedly different enzyme profiles. This finding suggests there may be a rather large variation in biologic response to the same hepatotrophic agent. None of the four chimpanzees had serologic evidence of acute infection with HAV, HBV, CMV,

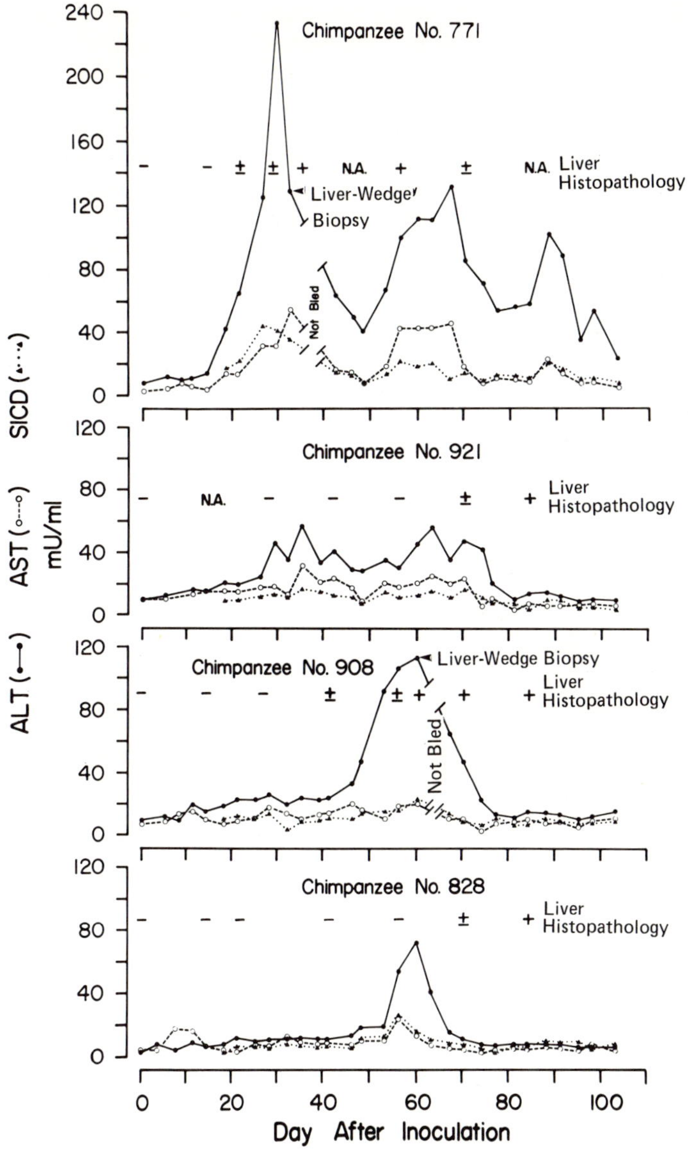

Figure 1 Serum enzyme profiles of four chimpanzees experimentally infected with implicated factor VIII (anti-hemophilic) materials. ALT = alanine aminotransferase; AST = aspartate aminotransferase; SICD = serum isocitrate dehydrogenase. (From D. W. Bradley, et al., *Journal of Medical Virology* 3:258, 1979.)

or EBV, and all showed histopathologic evidence of acute viral hepatitis in liver biopsy specimens. One chimpanzee (771) still had elevated ALT activity more than 2 years after inoculation and has subsequently been shown to be chronically infected.

Non A-non B hepatitis was subpassaged in four other chimpanzees by intravenous inoculation of either acute-phase plasma or liver obtained from chimpanzee 771, 29 and 32 days, respectively, after inoculation. Chimpanzees 509 and 770 (Fig. 2) exhibited distinctly different ALT profiles following inoculation of the same acute-phase plasma; however, their incubation periods (19 and 26 days) were similar. Inoculation of animals 824 and 799 with an acute-phase liver homogenate fraction resulted in elevated ALT activity 14 and 17 days, respectively, after inoculation (Fig. 3). The ALT elevations in these latter animals were more pronounced than in the animals inoculated with acute-phase plasma. As in the donor chimpanzee, there was no serologic evidence that HAV, HBV, CMV, or EBV was responsible for the acute hepatitis seen in these four chimpanzees. Liver histopathology indicative of viral hepatitis was mild in all four animals.

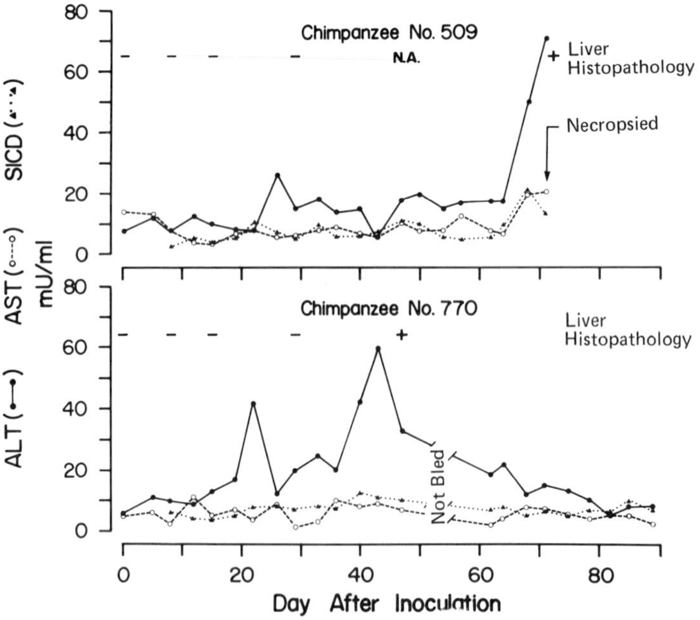

Figure 2 Serum enzyme profiles in two chimpanzees inoculated with acute-phase plasma from chimpanzee 771 (see Fig. 1). (From D. W. Bradley, et al., *Journal of Medical Virology* 3:260, 1979.)

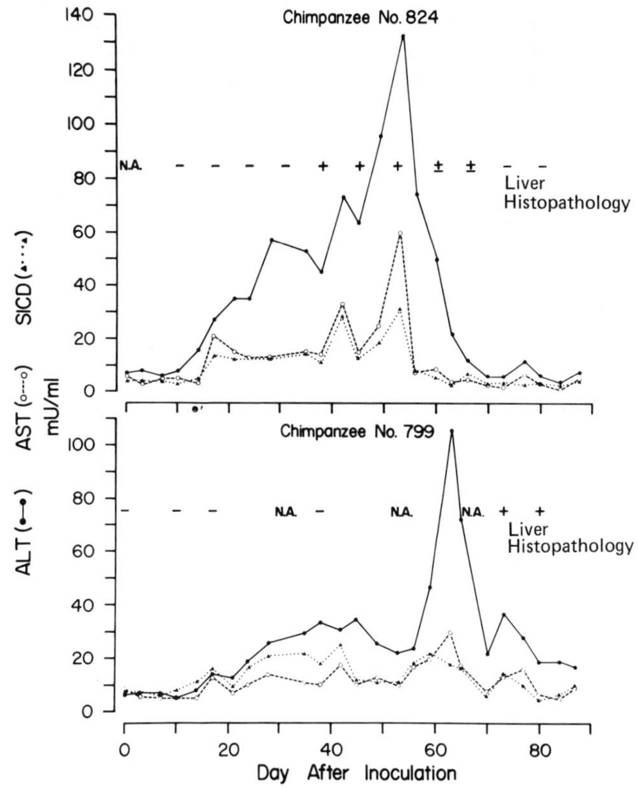

Figure 3 Serum enzyme profiles in two chimpanzees inoculated with an acute-phase liver preparation from chimpanzee 771. (From D. W. Bradley, et al., *Journal of Medical Virology 3*:260, 1979.)

Transmission of human-origin non A-non B hepatitis to chimpanzees by intravenous infusion of reconstituted, lyophilized factor VIII and IX materials experimentally verified that (1) these materials contained an infectious non A-non B hepatitis agent, and (2) the agent was capable of surviving lyophilization.

Marmosets

Preliminary studies in marmosets (*Saguinus mystax* and *S. labiatus*) suggest these subhuman primates may also be susceptible to infection by human-origin non A-non B hepatitis agent(s). One published study, however, showed that marmosets inoculated with acute-phase non A-non B hepatitis sera did not demonstrate biochemical (enzymatic) evidence of hepatitis. This study notwithstanding, other investigators have accumulated evidence that show inoculated marmosets do exhibit significant elevations in serum isocitrate dehydrogenase (SICD) or ALT activity indicative of acute hepatitis, with incubation periods ranging between 8 days and 22 weeks. Inocula have included human acute-phase and chronic-phase sera, as well as chimpanzee acute-phase sera and liver homogenates. Unfortunately, insufficient data are presently available to evaluate the usefulness of the marmoset as a biomedical model for the study of human non A-non B hepatitis.

DETECTION OF NON A-NON B HEPATITIS-ASSOCIATED ANTIGENS, ANTIBODIES, AND VIRUS-LIKE PARTICLES

Serologic Studies

Shortly after the reported transmission of human-origin non A-non B hepatitis to chimpanzees, several investigators presented evidence for the development of specific tests for non A-non B hepatitis antigen and/or antibody. The first of these reports described a double immunodiffusion assay for hepatitis "C" antigen in acute-phase sera of patients with non A-non B posttransfusion hepatitis. The antigen was distinct from HAV and HBsAg, migrated (under immunoelectrophoresis) in the serum β-globulin region, possessed a buoyant density of 1.30 g/cm^3 in cesium chloride, and had a molecular weight of between 100,000 and 300,000 daltons (d). A subsequent report from another group of investigators also showed evidence for the development of a similar immunodiffusion assay for a specific non A-non B hepatitis-associated antigen in acute-phase sera or in liver extracts from patients with chronic active non A-non B hepatitis. The antisera used in these latter studies were from a repeatedly transfused hemophiliac, a nurse convalescent from acute non A-non B hepatitis, and a chimpanzee that had been "boosted" with an acute-phase liver homogenate (from another chimpanzee) following apparent resolution of experimentally induced non A-non B hepatitis. These same workers also described a direct immunofluorescent assay for a non A-non B hepatitis-specific antigen in liver sections of patients with non A-non B hepatitis. Using fluorescein isothiocyanate (FITC)-labeled IgGs purified from the three convalescent sera, these investigators observed nuclear fluorescence in liver biopsies from all four patients with non A-non B hepatitis. Other worders also reported the detection of nuclear fluorescence in liver biopsies that appeared to be specifically associated with non A-non B hepatitis. The nuclear antigen was detected by an indirect immunofluorescent assay in which acute-phase chimpanzee liver biopsies were incubated sequentially with a convalescent-phase chimpanzee serum and a commercially prepared, FITC-labeled horse

IgG anti-human globulin. Nuclear antigen was found in liver biopsies from all chimpanzees that had been experimentally infected with either human serum or an acute-phase non A-non B hepatitis chimpanzee serum. Nuclear antigen was detected in liver biopsies obtained from nine chimpanzees between 2 and 30 weeks after inoculation, as well as in preinoculation liver biopsies from a tenth animal. Liver biopsies from other uninoculated chimpanzees, chimpanzees with HBsAg, or chimpanzees with acute viral hepatitis A, however, did not show nuclear fluorescence. These same workers also reported the detection of an antigen-antibody system by counterimmunoelectrophoresis (CEP) that was associated with human non A-non B hepatitis. CEP was used to detect an antigen in six of seven chimpanzees during the acute phase of experimentally induced non A-non B hepatitis. The antiserum was from a chimpanzee convalescent from human-origin non A-non B hepatitis. No similar antigen was detected in 35 preinoculation sera, nor was it found in 94 weekly bleedings from three chimpanzees with hepatitis A or three chimpanzees with hepatitis B. This antigen, however, was detected in serum samples obtained from three patients with chronic non A-non B hepatitis. Homologous antibody was reportedly detected in convalescent sera from all seven of the above non A-non B hepatitis-infected chimpanzees, in convalescent serum from a nurse with non A-non B hepatitis, as well as in serum from a hemodialysis patient recovered from non A-non B hepatitis.

The accumulated data from the serologic tests described above initially suggested that serodiagnosis of acute and chronic non A-non B hepatitis would be possible by the appropriate manipulation of rather simple and somewhat insensitive assays. Attempts to verify the specificity of the double-immunodiffusion, CEP, direct immunofluorescence, and indirect immunofluorescence assays for either a non A-non B hepatitis antigen or antibody, however, have met with frustration and failure in most other laboratories. Several groups of workers subsequently concluded there were a variety of nonspecific antigens and antibodies that could also be detected by these tests in sera or liver from patients or chimpanzees infected with non A-non B hepatitis. In fact, convalescent sera from many non A-non B hepatitis-infected patients and chimpanzees have recently been shown to contain antibodies against normal serum proteins and cellular components. For example, a unique "antigen" detected by RIA in the acute-phase serum of a chimpanzee infected with a factor VIII product was later found to be an IgM antibody specific for human (and not chimpanzee) IgG. Since the RIA employed IgGs prepared from presumed convalescent human serum, chimpanzee IgM anti-human IgG antibody contained in the acute-phase serum behaved (predictably) as an "antigen" in the above test.

For the present, it is generally agreed there is no specific test for any non A-non B hepatitis-associated antigen(s) or antibody. In spite of intensive laboratory efforts to identify a bona fide antigen or anti-non A-non B antibody, there have been no unequivocal results by immunodiffusion, CEP, immunofluorescence, enzyme immunoassay, or RIA. The lack of a detectable antigen or antibody may be explained by one or more of the following conditions: (1) low concentration of virus (antigen), (2) lack of a potent convalescent antibody, (3) complexes of viral antigen and specific antibody, and (4) inherent instability of antigen. Recent studies, in fact, have shown that a high percentage of experimentally infected chimpanzees develop persistent non A-non B hepatitis with viremia and elevated ALT activity. In view of these findings, several workers have speculated that many experimentally infected chimpanzees may not develop anti-non A-non B antibody.

Recovery of Virus-like Particles

Sera, plasma, and liver extracts from humans and chimpanzees with non A-non B hepatitis have been examined by several investigators for the presence of virus-like particles (VLPs) specifically associated with non A-non B hepatitis. In one study, sera from 28 patients with acute or chronic non A-non B hepatitis were examined by direct electron microscopy or immune electron microscopy (IEM) for VLPs after concentration by ultrafiltration. Twelve sera were found by EM to contain HBV-like particulate structures with diameters of 15 to 25 nm (small spherical particles and tubules) and 35 to 40 nm (double-shelled particles with an electron-dense core). Of these sera, 10 were from patients with a non A-non B antigen detectable by double immunodiffusion (see above discussion) and 2 were from patients without detectable antigen. IEM examination of these same sera revealed antibody-aggregated HBV-like structures when anti-non A-non B antiserum was used; however, similar aggregates were not observed in any serum reacted with anti-HBs. These same workers reported the recovery of HBV core-like particles from two fractionated human liver homogenates. The autopsy livers were obtained from two patients who had presumably died of non A-non B hepatitis. Endogenous DNA polymerase activity was also found associated with the partially purified core-like structures. Unfortunately, the finding of HBV-like particles in the sera of non A-non B hepatitis patients has not been verifiable in other laboratories, nor have similar particles been recovered from the acute-phase sera or plasma from chimpanzees that have been experimentally infected with a wide variety of non A-non B hepatitis materials.

Distinctly different virus-like particles were reportedly recovered from the serum of a hemodialysis patient with non A-non B hepatitis, as well as from concentrated urine samples from two icteric non A-non B hepatitis cases. Electron microscopic examination of these specimens revealed 60 nm diameter particles with 40 nm cores. The purported VLPs did not appear to be aggregated in clusters of more than two particles each, as indicated by the published electron micrographs. Other investigators have not visualized similar VLPs in acute-phase non A-non B hepatitis serum or liver specimens.

In a separate study, conducted in 1978, distinct 27 nm diameter virus-like particles were visualized by immune electron microscopy in a factor VIII product that was shown to induce non A-non B hepatitis in five chimpanzees. The VLPs varied in diameter from 25 to 30 nm (average 27 nm) and were somewhat ragged in appearance. All the particles were coated with antibody, presumably from the antiserum used in the IEM assay. Most of the particles were hollow in appearance. No other VLPs with a different morphology were visualized in the factor VIII material. Intravenous inoculation of a chimpanzee (771, described above) with this factor VIII product resulted in short-incubation non A-non B hepatitis. An open liver-wedge biopsy was obtained from this animal 32 days after inoculation when ALT activity was near its peak value. The liver was homogenized and then banded in a CsCl density gradient. The gradient fractions were examined by IEM for VLPs using presumed convalescent serum from a patient who had contracted non A-non B hepatitis following infusion of the same factor VIII material. Aggregates of antibody-coated 27 nm VLPs were detected primarily in two gradient fractions with a mean buoyant density of 1.31 g/cm^3 (Fig. 4). These particles were morphologically similar to those visualized in the inoculum. Both electron-dense (full) and electron-luscent (empty) VLPs were found within the same aggregate. Many, if not most, of the VLPs were aggregated by IgM (acute-phase) antibody. Some of the aggregates contained VLPs that appeared to be ragged or distorted; the majority of the empty particles (Fig. 5) were somewhat degraded. The latter particles were most typical of those visualized in the CsCl gradient fractions. Electron microscopic examination of the peak CsCl gradient fractions using buffer or control sera (that is, sera from chimpanzees or humans not previously exposed to non A-non B hepatitis) revealed the presence of smaller numbers of antibody-coated VLPs. These particles were morphologically indistinguishable from those visualized by IEM using "convalescent" serum from the index patient described above.

Intravenous inoculation of another chimpanzee with CsCl gradient fractions of the first-passage chimpanzee liver containing the 27 nm VLPs induced elevated ALT activity approximately 2 weeks later. An open liver-wedge biopsy was obtained from this animal 13 days after inoculation, or approximately 1 week before the peak of ALT activity. CsCl gradient fractions from an isopycnic banding of the second-passage chimpanzee liver extract were examined by IEM for VLPs using

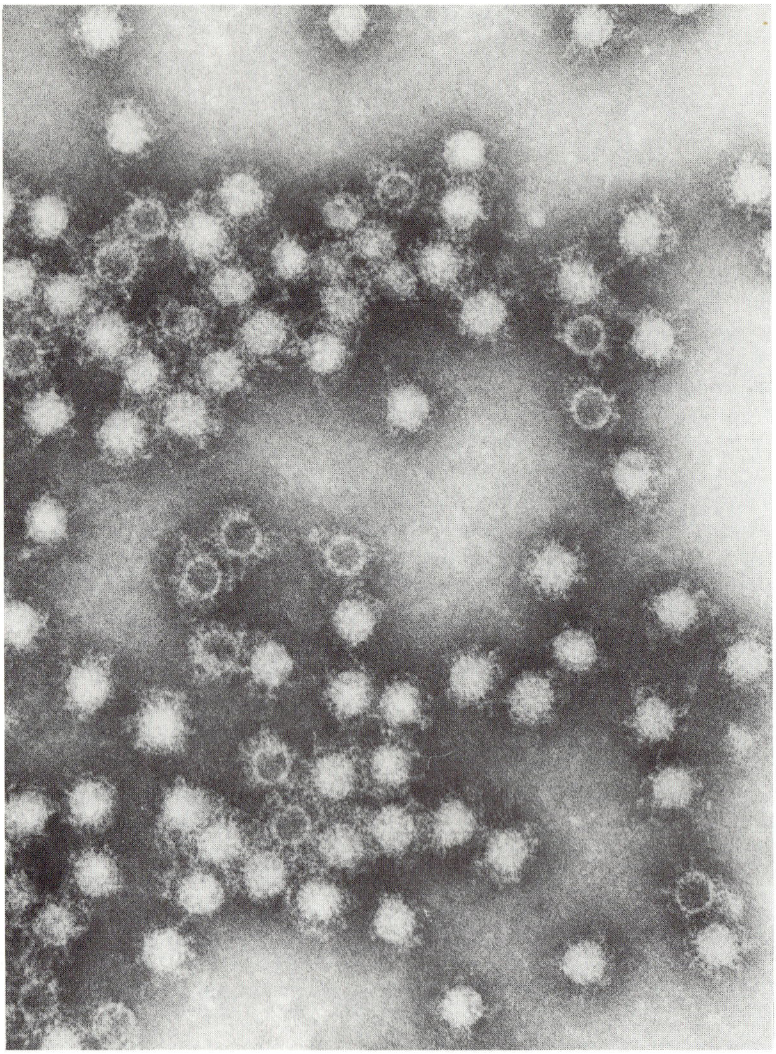

Figure 4 Twenty-seven nanometer virus-like particles recovered from an acute-phase chimpanzee liver homogenate (animal 771). Virus-like particles are shown aggregated by antibody. Magnification 206,700X. (From D. W. Bradley, et al., *Journal of Medical Virology* 3:263, 1979.)

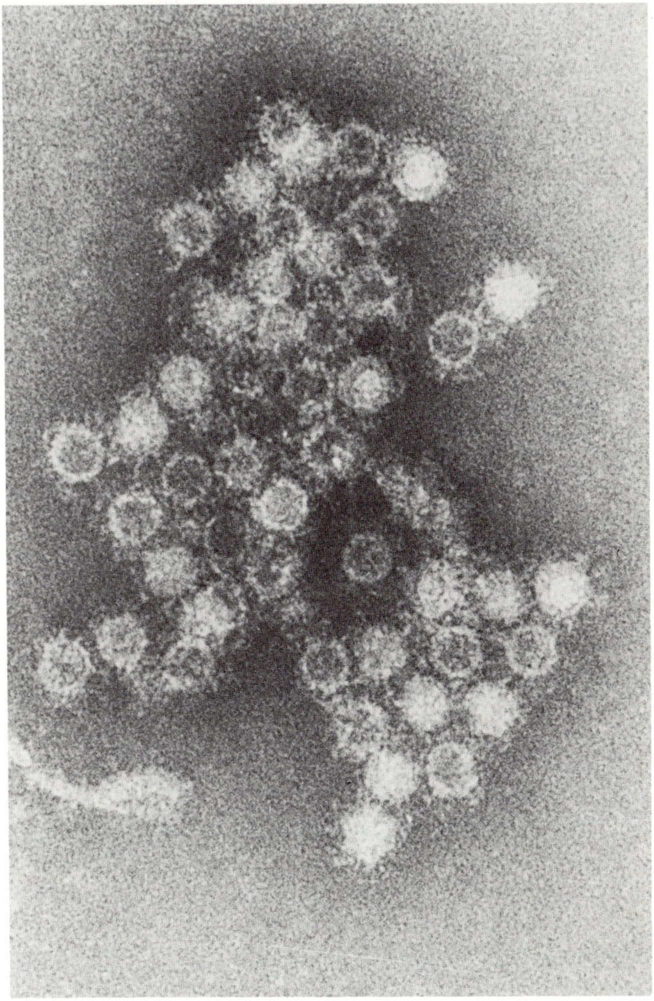

Figure 5 Twenty-seven nanometer virus-like particles recovered from a first-passage chimpanzee (771) liver homogenate banded on a cesium chloride density gradient. Magnification 246,900X.

a pool of three presumed convalescent non A-non B hepatitis sera as the source of antibody. Aggregates of empty and full antibody-coated VLPs were visualized primarily at a buoyant density of 1.30 g/cm^3 (Fig. 6). The particles ranged in diameter from 25 to 32 nm (average 27 nm). Other aggregates of similarly-sized VLPs were also visualized in these same gradient fractions. Many of these particles were partially obscured by background material, most likely cellular membranes, or excess antibody. Some of the VLPs also appeared to be coated by IgM antibody. As in the other liver, no VLPs with a size and morphology different from those described above were found by IEM in any of the gradient fractions.

Acute-illness phase sera from patients with non A-non B hepatitis were also examined by IEM for the presence of VLPs. Only 2 of 20 sera examined contained well-defined 27 nm diameter VLPs. These particles were of the same size and morphology as those found in the factor VIII inoculum, first-passage chimpanzee liver, and second-passage chimpanzee liver. No other VLPs with a different size and morphology, including HBV-like particles, were detected by IEM in any of the acute-phase sera. No distinct VLPs were detected by IEM in any of 14 sera from healthy individuals with normal ALT activity and no history of hepatitis.

Intravenous inoculation of four marmosets with the first-passage chimpanzee liver and three marmosets with the second-passage chimpanzee-liver preparation described above did not cause sero-conversion to HAV, indicating the 27 nm VLPs recovered from these livers were unrelated to HAV. Also, immunofluorescent assays for HAV, HBsAg, and HBcAg in several of the chimpanzee acute-phase liver biopsies were negative and further excluded the possibility that the 27 nm VLPs visualized in the same acute-phase non A-non B hepatitis livers were either HAV or HBV core particles.

The finding of 27 nm diameter virus-like particles associated with non A-non B hepatitis was confirmed approximately 1 year later when another group of workers visualized morphologically identical 27 nm VLPs in (1) a plasma fraction (fibrinogen) implicated in the transmission of non A-non B hepatitis, (2) the acute-phase serum of two chimpanzees inoculated with an implicated donor plasma, and (3) sera from 8 of 100 apparently healthy blood donors with ALT activity in excess of 80 Karmen units (40 mU) per milliliter of serum. Two patients contracted non A-non B hepatitis 3 and 8 weeks, respectively, after infusion of the implicated plasma fraction. A chimpanzee inoculated with the same preparation developed non A-non B hepatitis 11 weeks later. Preacute-phase serum from this chimpanzee induced non A-non B hepatitis in another intravenously inoculated chimpanzee after an incubation period of 8 weeks. Inoculation of two other chimpanzees with 5.0 ml of donor serum positive by IEM for 27 nm VLPs resulted

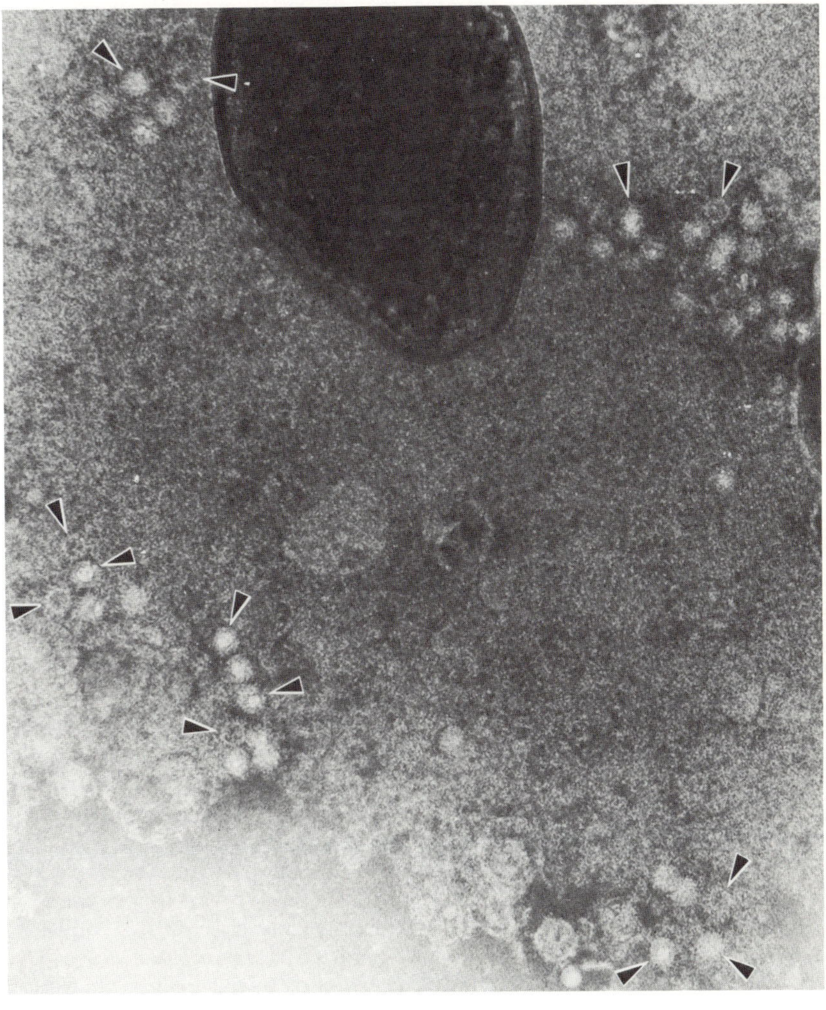

Figure 6 Twenty-seven nanometer virus-like particles recovered from a second-passage chimpanzee (825) liver homogenate banded on a cesium chloride density gradient. Magnification 139,000X.

in non A-non B hepatitis after an incubation period of approximately 6 weeks. Virus-like particles with a diameter of 27 nm and a morphology indistinguishable from those in the inoculum were observed by IEM in the necropsy sera of these chimpanzees obtained 6 and 6½ weeks, respectively, after inoculation. All the recovered VLPs were electron dense. Similar VLPs were not recovered from the preinoculation sera of these two chimpanzees, nor were the above particles aggregated by preinoculation sera from the same animals. Isopycnic banding of the chimpanzee serum in a CsCl gradient showed 27 nm VLPs banding primarily at a buoyant density of 1.28 g/cm^3, a buoyant density similar to that of the previously described 27 nm VLPs recovered from acute-phase chimpanzee liver homogenates.

PATHOGENESIS OF DISEASE

Clinical Characteristics

Acute non A-non B hepatitis is normally accompanied by mild clinical symptoms and a relatively low frequency of jaundice. The incubation period for non A-non B hepatitis in transfused patients, or in recipients of clotting factors, has been reported to range from 5 days to longer than 20 weeks. Elevations of ALT activity are generally less than those associated with acute viral hepatitis A or B; however, a high proportion of acutely infected individuals develop chronic hepatitis. Percutaneous liver biopsies from individuals with persistent enzyme elevations, for example, have often shown histopathologic evidence of either chronic active or persistent hepatitis.

Numerous reports of the progression of acute non A-non B posttransfusion hepatitis to chronic hepatitis have appeared in the literature in the past several years and deserve further comment. Several workers have shown that patients with hemophilia are at a high risk of developing acute or chronic non A-non B hepatitis as a result of repeated infusions of the clotting factors required for treatment. In a survey of 91 patients at one hemophilia center, 45% were found to have moderate elevations of either ALT or AST activity; however, less than 25% demonstrated abnormal levels of serum bilirubin. None of these patients showed clinical symptoms indicative of liver disease. Three other similar studies conducted in the United States showed the persistence of elevated ALT or AST activity in more than 40% of the patients tested. Percutaneous liver biopsies from four HBsAg-negative hemophiliac patients who had demonstrated abnormal transaminase activities for periods of time ranging between 7 and 24 months showed histopathologic evidence of CAH in one and CPH in three. All four patients had previously received factor VIII concentrates, whole blood, plasma, or cryoprecipitate. In another study, 36 (77%)

of 47 factor VIII-treated hemophiliacs were found to have elevated transaminases; two patients were positive for HBsAg. Of the 36 patients, 25 showed persistent enzyme elevations for more than 6 months; 8 of these patients were subsequently selected for percutaneous liver biopsy. All 8 liver biopsies showed evidence of CAH, CPH, or cirrhosis. Markers of HBV infection were absent in the sera of 4 of these patients, and only 2 sera from the remaining 4 patients were positive for HBsAg. These combined findings suggest that chronic active and chronic persistent hepatitis are common sequalae in hemophiliacs who have presumably contracted non A-non B hepatitis following infusion of one or more types of blood products.

Outbreaks of non A-non B hepatitis have been documented in hemodialysis centers in the United States and in Europe. Reports of two outbreaks showed that chronic liver disease often followed acute non A-non B hepatitis. Liver biopsies from seven of eight patients who had demonstrated elevated transaminase activity for 4 years or more following one outbreak of non A-non B hepatitis showed CAH in three and CPH in two. The remaining two biopsies had nonspecific hepatitis in association with massive iron overload (grade IV siderosis).

Chronic liver disease following acute non A-non B posttransfusion hepatitis has been documented by several groups of workers. In one study, 44 cases of acute non A-non B posttransfusion hepatitis were prospectively followed for the development of chronic liver disease. Ten patients (23%) demonstrated elevated transaminase activities for 12 to 36 months after the acute phase of disease. Liver biopsies from these patients showed histopathologic evidence of CAH (9), CPH (1), or cirrhosis (1). The authors concluded that acute non A-non B posttransfusion hepatitis often progresses to chronic active hepatitis. Other workers have analyzed the frequency of chronic liver disease in three epidemiologically distinct populations experiencing acute non A-non B hepatitis, including recipients of transfusions, illicit drug users, and sporadic cases with no obvious exposure. Of the 45 cases of acute non A-non B hepatitis, 18 (40%) developed chronic liver disease. Chronic liver disease was observed in 7 (54%) of the 13 individuals with posttransfusion hepatitis and in 7 (58%) of the 12 cases who were addicts. Chronic liver disease, however, developed in only 4 (20%) of the acute cases who had no obvious exposure. Chronic-phase liver biopsies obtained from 6 of the 7 cases of posttransfusion hepatitis showed either CAH (4) or CPH (2). Chronic-phase liver biopsies obtained from 5 of the drug addicts showed CPH; CPH was also demonstrable in 3 chronic-phase liver biopsies obtained from 4 of the cases with no obvious exposure.

In a major prospective study of 388 patients followed after open-heart surgery, 26 (6.7%) were found to develop non A-non B hepatitis after transfusion of volunteer donor blood. Of these latter patients,

12 (46%) demonstrated persistent or intermittent elevations in ALT activity for more than 1 year. Liver biopsies from 8 of these 12 patients showed histopathologic evidence of CAH in 6 and CPH in 2. Chronic liver disease (CLD) more often developed in anicteric patients whose peak ALT activity was greater than 300 mU/ml than in anicteric patients whose peak enzyme value was less than 300 mU/ml. None of the patients with CLD developed anti-smooth muscle or anti-mitochondrial antibodies. ALT activity in all 12 patients with CLD showed significant improvement over periods of time ranging from 1 to 3 years; ALT activity in 4 of these patients eventually returned to normal values.

Electron Microscopic Studies

To date, no electron microscopic studies of liver biopsy materials obtained from humans with acute or chronic non A-non B hepatitis have been reported. However, several groups of investigators have described ultrastructural changes in the hepatocytes of acutely infected chimpanzees. In one of the first reported studies, acute-phase liver biopsies were obtained from nine experimentally infected chimpanzees. Of these animals, four had been inoculated with plasma from a patient with chronic non A-non B posttransfusion hepatitis (plasma F), and five had been inoculated with a second plasma obtained from a patient with acute non A-non B hepatitis (plasma H). All four animals inoculated with the F plasma developed non A-non B hepatitis with elevated ALT activity, while only two of the chimpanzees inoculated with aliquots of the H plasma developed elevated ALT activity. Electron microscopic examination of acute-phase liver biopsies obtained from the F plasma-inoculated chimpanzees revealed the presence of unique tubular structures in the hepatocyte cytoplasm (Fig. 7). In thin section, these tubular structures appeared circular in cross-section and composed of parallel membranes when cut longitudinally. The tubule walls were approximately 25 nm in thickness and were comprised of two unit membranes enclosing electron-dense material; the endoplasmic reticulum sometimes appeared to be contiguous with the outer membranes. The total diameter of the tubular structures ranged between 150 and 300 nm. The lengths of the tubular structures also varied, with the longest being 2200 nm; no nuclear alterations were observed in these hepatocytes. EM examination of liver biopsies from the H plasma-inoculated chimpanzees did not show the characteristic hepatocyte cytoplasmic structures described above. However, nuclear alterations were observed in the hepatocytes of these animals, including crenellation of the nuclear membranes and margination of the chromatin. Intranuclear aggregates of 20 to 27 nm particles were visualized in the hepatocytes of four of the five chimpanzees; however, the particles were not as uniform as small viruses normally appear. On the basis

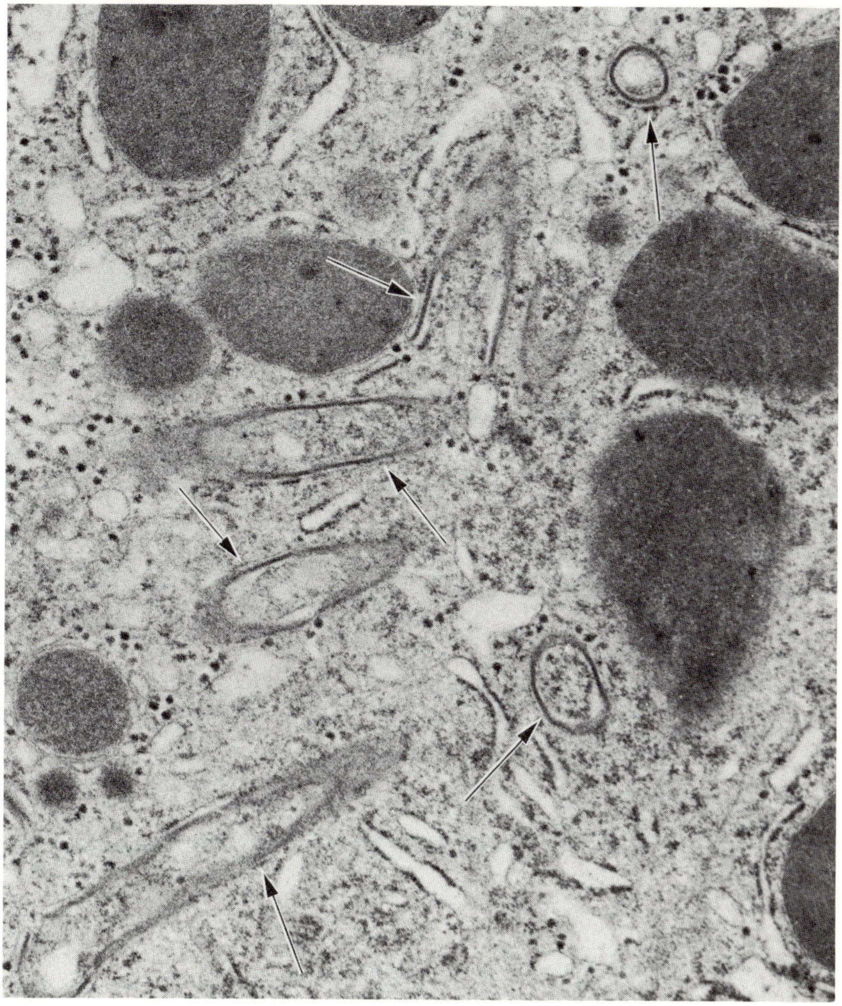

Figure 7 Unique tubular structures found in the cytoplasm of a hepatocyte from a chimpanzee infected with the factor VIII non A-non B hepatitis agent. Magnification 29,600X. (From D. W. Bradley, et al., *Journal of Medical Virology* 6:193, 1980.)

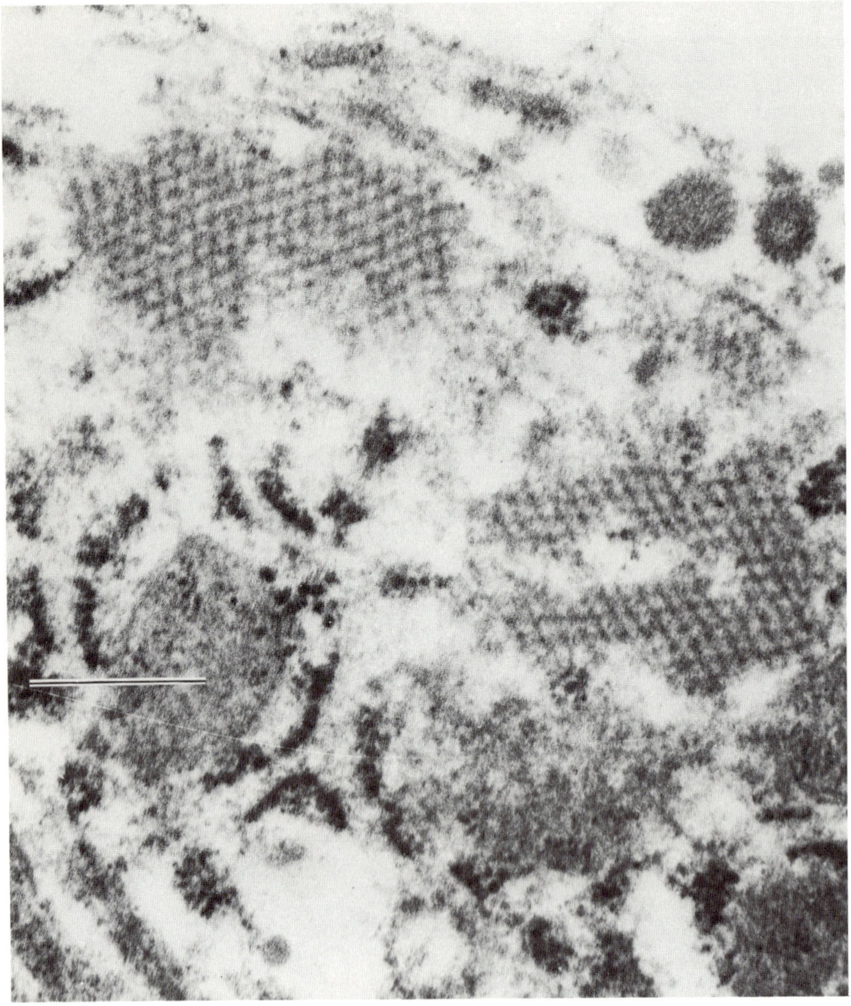

Figure 8 Crystalline array of 25 to 34 nm particles in an acute-phase liver biopsy (endothelial cells). Crystalline arrays were detected by EM in acute-phase liver cells from a total of four chimpanzees, three inoculated with the factor VIII agent and one inoculated with plasma H identified in the text. Magnification 79,300X. (From D. W. Bradley, et al., *Journal of Medical Virology* 6:197, 1980.)

of these findings, these workers suggested the apparently mutually exclusive nuclear and cytoplasmic ultrastructural alterations might have been caused by two different etiologic agents of non A-non B hepatitis.

Other investigators have also reported the finding of unique ultrastructural alterations in the cytoplasm of infected hepatocytes. Acute-phase liver biopsies obtained from five chimpanzees infected with sera from three patients with chronic non A-non B showed a characteristic abnormal derangement of the hepatocyte endoplasmic reticulum in four of the animals. These changes were identical to those described above and consisted of tubules that appeared to be randomly distributed in the cytoplasm. These cytoplasmic tubules were not visualized by EM in acute-phase liver biopsies from chimpanzees infected with viral hepatitis A or B.

In another preliminary study, two distinct types of ultrastructural changes were found in the hepatocytes of chimpanzees infected with two presumed agents of non A-non B hepatitis. Chimpanzees infected with the factor IX inoculum (described above) showed the characteristic hepatocyte cytoplasmic tubules in acute-phase liver biopsies. The nuclei in these hepatocytes appeared normal. Acute-phase liver biopsies from a chimpanzee infected with the factor VIII inoculum, however, were initially reported to show nuclear alterations similar to those induced in chimpanzees by plasma H. No cytoplasmic changes were observed in acute-phase hepatocytes. Subsequent reexamination of these liver biopsies and acute-phase liver biopsies obtained from another factor VIII-inoculated chimpanzee revealed the presence of the characteristic cytoplasmic tubular structures. These findings suggested that the tubular structures were more specifically associated with acute non A-non B hepatitis than the nuclear alterations and that infection produced by two presumed agents of non A-non B hepatitis could not be differentiated on the basis of the attendant ultrastructural alterations. Other investigators have also recently reported the finding of tubular cytoplasmic structures in acute-phase liver biopsies obtained from chimpanzees infected with factor VIII, plasma H, or a plasma fraction (fibrinogen). No nuclear alterations were consistently observed by EM in infected hepatocytes.

Some investigators have also reported finding crystalline arrays of particles in the cytoplasm of endothelial or Kupffer cells (Fig. 8). The particles measured approximately 25 to 34 nm in diameter, but did not possess the normally distinct morphology of small virus particles observed in crystalline lattices. These crystalline arrays were visualized in an acute-phase chimpanzee liver biopsy from which 27 nm diameter virus-like particles had been previously recovered. Additional studies will be needed, however, to determine the relationship of these crystalline arrays to the reported 27 nm VLPs.

SUGGESTED READINGS

Alter, H. J., R. H. Purcell, P. V. Holland, and H. Popper. Transmissable agent in non-A, non-B hepatitis. *Lancet 1*:459-463, 1978.

Berman, M., H. J. Alter, K. G. Ishak, R. H. Purcell, and E. A. Jones. The chronic sequelae of non-A, non-B hepatitis. *Ann. Intern. Med. 91*:1-6, 1979.

Bradley, D. W., E. H. Cook, J. E. Maynard, K. A. McCaustland, J. W. Ebert, G. H. Dolana, R. A. Petzel, R. J. Kantor, A. Heilbrunn, H. A. Fields, and B. L. Murphy. Experimental infection of chimpanzees with antihemophilic (factor VIII) materials: Recovery of virus-like particles associated with non-A, non-B hepatitis. *J. Med. Virol. 3*:253-269, 1979.

Francis, D. P., and J. E. Maynard. The transmission and outcome of hepatitis A, B, and non-A, non-B: A review. *Epidemiol. Rev. 1*:17-31, 1979.

Jackson, D., E. Tabor, and R. J. Gerety. Acute non-A, non-B hepatitis: Specific ultrastructural alterations in endoplasmic reticulum of infected hepatocytes. *Lancet 1*:1249-1250, 1979.

Purcell, R. H., H. J. Alter, and J. L. Dienstag. Non-A, non-B hepatitis. *Yale J. Biol. Med. 49*:243-250, 1976.

Tabor, E., R. J. Gerety, J. A. Drucker, L. B. Seeff, J. H. Hoofnagle, D. R. Jackson, M. April, L. F. Barker, and G. Pineda-Tamondong. Transmission of non-A, non-B hepatitis from man to chimpanzee. *Lancet 1*:463-466, 1978.

Yoshizawa, H., Y. Akahane, Y. Itoh, S. Iwakiri, K. Kitajima, M. Morita, A. Tanaka, T. Nojiri, M. Shimizu, Y. Miyakawa, and M. Mayumi. Viruslike particles in a plasma fraction (fibrinogen) and in the circulation of apparently healthy blood donors capable of inducing non-A/non-B hepatitis in humans and chimpanzees. *Gastroenterology 79*:512-520, 1980.

chapter 6
The Unconventional Viruses

DAVID T. KINGSBURY

College of Medicine
University of California, Irvine
Irvine, California

As our awareness of the epidemiology and etiology of many chronic diseases of the central nervous system grows we begin to realize the complexity of the diseases referred to as "slow" infections. The concept of the slow infection was first articulated by Sigurdsson in 1954 as a result of his studies of diseases of sheep in Iceland. Sigurdsson's criteria for a slow infection were: (1) a prolonged incubation period measurable in months or years, (2) a regular clinical course of the infection progressing to serious disease or death, (3) pathology limited to a single organ system, and (4) limitation of the disease to a single species of host. These criteria, based on the study of scrapie, visna, and maedi, still serve to outline the nature of slow infections. Since the original description it has become clear that slow infections of the central nervous system are associated with two broadly classified groups of agents. First, we recognize that viruses with conventional properties, most of them associated with acute illness, under the right conditions, can cause a chronic and protracted disease course. An example of such an occurrence is the association of measles (rubeola) virus with a progressive central nervous system disease, subacute sclerosing panencephalitis (SSPE). The SSPE virus is clearly not a normal measles virus but is most likely a mutant, defective, or recombinant form of the virus. In addition to chronic infection by the conventional viruses, slow infections are also associated with agents with "unconventional" properties. The most well studied of these is scrapie of sheep and goats.

Scrapie has been recognized as a naturally occurring disease in flocks of sheep and goats for several hundred years. It is striking in several features, one of which is the noninflammatory nature of the pathology. Scrapie serves as a model of what are now termed the subacute spongiform virus encephalopathies. There are four generally recognized diseases of this type, scrapie and transmissible mink encephalopathy of animals and two human diseases, kuru and Creutzfeldt-Jakob disease (CJD).

BIOLOGIC AND BIOCHEMICAL PROPERTIES OF THE UNCONVENTIONAL VIRUSES

The biologic and biochemical properties of the unconventional viruses have proven to be elusive and conflicting. The epidemiology of the natural diseases is understood only in the case of kuru. This viral group presents the virologist with an awesome challenge, to both modern technology and biologic concepts. The characterization of the biologic and biochemical properties of the unconventional viruses has been confined almost exclusively to the agent of scrapie. The majority of this work has been done in the experimental hosts, the mouse and hamster. The absence of an in vitro or convenient in vivo assay system has almost completely thwarted the development of our knowledge of this group of agents. Kuru and Creutzfeldt-Jakob disease are almost entirely confined to primates as experimental hosts, and only recently has CJD been adapted to the guinea pig and mouse. It is probably the case that the unconventionality of these agents stems from the absence of "conventional wisdom" about their properties and transmission.

To be sure, many of the physical and chemical properties of the unconventional viruses are unique to this biologic group. The agents are unusually resistant to numerous chemical and physical treatments. These are listed in Table 1. The most striking of the chemical resistances are to formaldehyde, glutaraldehyde, and β-propiolactone. The other properties listed, for example resistance to heat, are not unique to this group. However, when taken together with their other unusual properties they present a picture of a biologic group distinct from any that has been previously described. The unconventional viruses exhibit an unusually high resistance to both ultraviolet (UV) and ionizing radiation. The action spectrum for scrapie inactivation is also highly unusual for viral agents, with a sixfold greater sensitivity at 230 nm over that between 254 and 280 nm. This unusual action spectrum has led some to postulate that the scrapie agent lacks nucleic acid. However this unusual resistance to nonionizing radiation is similar to that observed for the plant viroids (naked circular RNA molecules which cause a number of naturally occuring plant diseases,

Table 1 Physical and Chemical Resistances of the Unconventional

I.	Resistant to:
	Formaldehyde
	Glutaraldehyde
	β-Propiolactone
	EDTA
	Heat, little effect at 80°C, incomplete inactivation at 100°C
	Proteases (trypsin, pepsin, pronase, proteinase K)
	Nucleases (DNase, RNase, micrococcal nuclease)
	Phospholipases
	Nonionic or mildly ionic detergents (Triton, NP40, deoxycholate, Sarkosyl)
II.	Sensitive to:
	Autoclaving (121°C, 15 psi, 60 min)
	Phenol
	Sodium hypochlorite (5%)
	Ether and acetone (only moderate sensitivity)

such as potato spindle tuber disease) in crude preparations. These similarities have led to the common postulate that the unconventional viruses are animal viroids. Experiments utilizing cobalt-60 irradiation of the scrapie agent have more reliably revealed the target size as approximately 150,000 daltons in mass, smaller than any other known animal virus and representing nothing more than the genetic information necessary for a single small polypeptide.

In addition to their biophysical properties the unconventional viruses also have a series of biologic properties unique among infectious agents. The most notable of these is the complete absence of an inflammatory response in infected animals. Other properties which are not totally unique to this group are their long incubation period, the chronic progressive nature of the pathologic lesions, and the absence of remission or recovery from infection, the diseases always being fatal. It is within the framework of these unusual properties that most studies on the biophysical properties of the agents have been focused.

PHYSICAL PROPERTIES OF THE UNCONVENTIONAL AGENTS

The physical properties of the scrapie agent isolated from the mouse or hamster have been studied by a number of investigators in various parts of the world. The results of these studies, although conflicting in many cases, have drawn a general picture of the agent and have shown that the scrapie agent behaves identically whether from a mouse, hamster, or goat. It has also been clearly demonstrated that the scrapie agent prepared from the spleen of mice early in infection is very similar or identical to that prepared from the brain much later in infection, although the titers achieved in the brain are much higher. To be sure, the brain material offers the investigator a much richer source of infectious virus to work with, although, because of the very nature of the brain tissue, the material is much more difficult to handle.

To date no virus-like structure has been clearly associated with the infectivity. Several investigators have described small osmiophilic particles of variable size but averaging approximately 35 nm in brain tissue. These particles are observed primarily in postsynaptic intradendritic processes and are associated with the regions of the brain exhibiting the most obvious pathologic changes. However, their presence is not common to all infected species. No such particles have ever been seen in scrapie-infected hamsters despite the very high levels of infectious material in these animals. These particles have been seen in several different experimental and natural infections by the unconventional viruses, but their identity has remained a mystery. No other type of unique structure has been observed in thin sections, negative-stained preparations, or by freeze fracture of infected tissue.

The picture that has emerged from the biophysical studies of the scrapie agent is unlike that seen with any other viral agent. It is clear that most, if not all, of the infectivity is associated with the cellular membranes, the majority being associated with the cytoplasmic membrane. The estimates of the membrane-associated infectivity range from 90 to 100%, depending on the particular study. This range is partially dependent upon the method of preparation of the homogenates and probably represents the fragmentation of the membranes into small particles which sediment with the soluble portion of the cell. The experience with the CJD agent has likewise confirmed that nearly 100% of the infectivity is associated with the cellular membranes. Figures 1 and 2 illustrate the behavior of scrapie and CJD infectivity when examined in centrifugation experiments in fixed-angle rotors. It is quite clear that the infectivity follows the sedimentation pattern of the cytoplasmic membrane as determined by cytoplasmic membrane enzyme markers assayed in this experiment. The inclusion of the membrane-disrupting agent deoxycholate in these preparations modifies this sedimentation pattern and solubilizes the cytoplasmic membrane marker enzymes, causing their behavior to resemble that of the markers for the soluble cytoplasmic enzymes. Likewise, accompanying the solubilization of the membrane is a dramatic change in the sedimentation

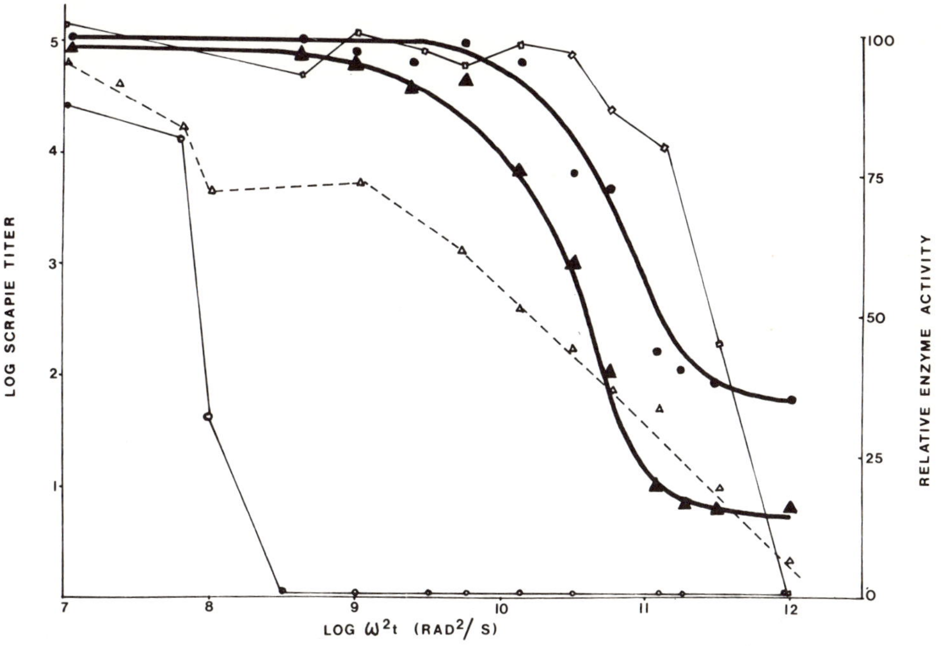

Figure 1 Sedimentation profiles of scrapie infectivity and cell organelle marker enzymes. The scrapie infectivity is represented by the heavy solid lines; (●) scrapie infectivity in untreated 10% mouse spleen homogenate treated with 0.5% deoxycholate. Succinate dehydrogenase (○) was used as a mitochondrial marker, acid phosphatase (△) as a cytoplasmic membrane marker, and lactate dehydrogenase (□) as a soluble enzyme marker.

profile of the scrapie and CJD agents. In both these cases, treatment of the preparations with deoxycholate modified the apparent sedimentation values from 625S in untreated membrane preparations to less than 150S following deoxycholate disruption. In fact, carefully quantitated sedimentation experiments using particulate viral sedimentation markers suggest that treatment of scrapie-infected brain with deoxycholate reduces the apparent sedimentation value to less than 40S. That these structures are not the smallest elementary form of scrapie is illustrated by the fact that removal of the deoxycholate or the subjection of these preparations to mild heating causes massive aggregation of the particles and a resulting sedimentation value of greater than 1200S. This suggests that the form of the scrapie agent in deoxycholate-treated membranes is either as a particle very rich in lipid or that the

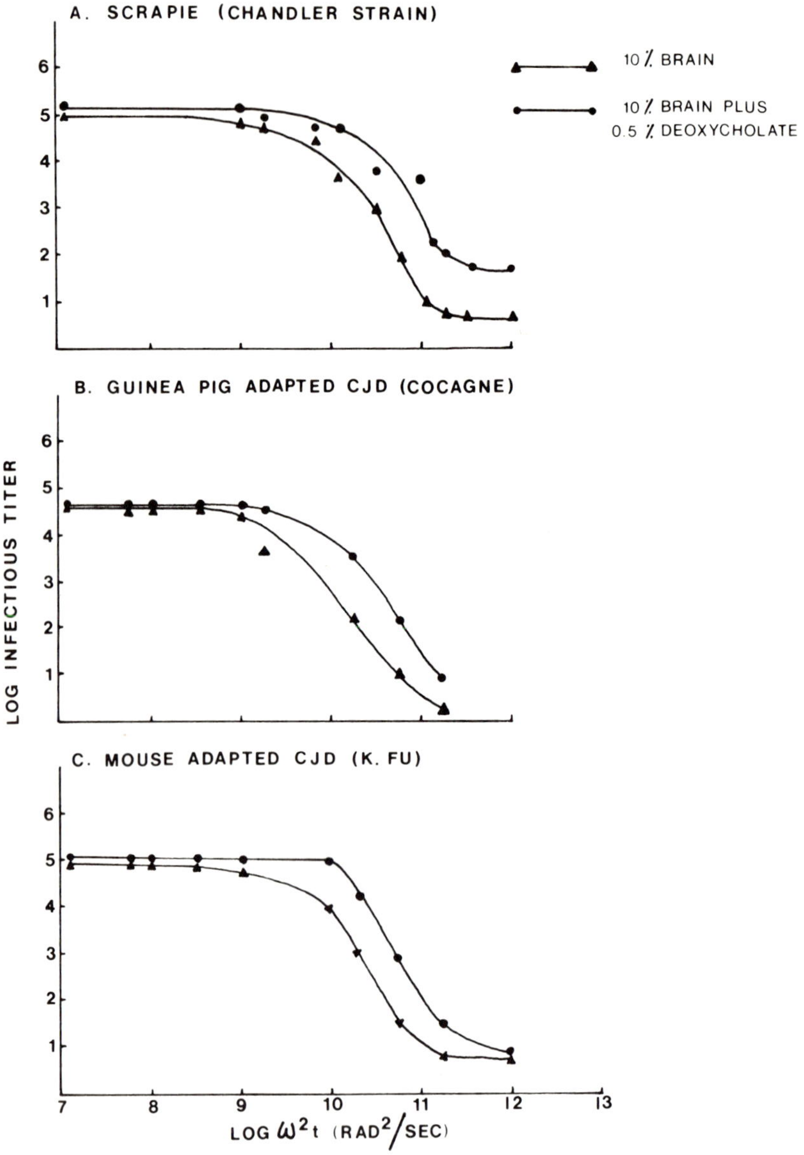

Figure 2 Comparative sedimentation profiles of untreated and 0.5% deoxycholate-treated brain suspensions.

particles derived in this way are simply small membrane fragments that have associated scrapie infectivity.

This detergent-mediated partial membrane disruption has formed the basis of the most successful purification schemes employed to date. Combining detergent treatment with precipitation or centrifugation procedures, a scrapie preparation purified greater than 20-fold with respect to cellular protein was characterized and found to contain a wide variety of different "molecular" (physical) forms of the scrapie agent. These multiple forms had densities which ranged from 1.08 to 1.30 g/cm^3. Likewise, the apparent sedimentation values in carefully determined experiments ranged from 40S to over 500S. Incubation of this partially purified preparation at 80°C induced a shift from low molecular weight forms to higher molecular weight forms, suggesting again that these molecules are covered with many hydrophobic domains.

It is clear that further purification of the scrapie agent is limited by the apparent physical heterogeneity. Gel electrophoresis of these preparations in the absence of any additional treatment likewise demonstrated a enormous degree of heterogeneity in the preparations. These scrapie preparations have now been more highly purified by a series of treatments with proteolytic and nucleolytic enzymes. The overnight digestion with micrococcal nuclease followed by proteinase K treatment has yielded a preparation with much more uniform electrophoretic migration properties, yet one that still has a high degree of apparent heterogeneity. It is interesting that preparations which constitute an approximate 1000- to 2000-fold purification over the starting brain suspension are still resistant to phospholipases A2 and C, DNases, RNases, and proteases of all types, as well as phosphodiesterases from snake venom. In addition, protease digestions were carried out in the presence of low concentrations of the detergents Sarkosyl (sodium dodecyl sarcosinate) or deoxycholate, all without effect. In contrast to the treatment with mild detergents, the agent of scrapie is sensitive to denaturation by the ionic detergent sodium dodecyl sulfate (SDS). This inactivation exhibits a cooperative binding effect, and denaturation occurs at an SDS-protein ratio of approximately 2:1, identical to that for many common enzymes. Electrophoresis of the agent into gels containing as little as 0.1% SDS leads to an almost complete loss of infectivity. When the scrapie preparations were applied to agarose gels (0.6%) containing 0.2% Sarkosyl, the agent was not dentatured but again behaved as a heterogeneous population. To be sure, under these conditions the heterogeneity of these preparations is dramatically reduced, and greater than one-third of the agent in the sample migrates through the sarkosyl agarose gel with a sedimentation value less than 40S. Thus, by both electrophoresis and sedimentation, the scrapie agent appears to be smaller than the smallest known conventional virus yet retains a high degree of heterogeneity and its hydrophobic behavior. The availability of these 1000+ fold purified preparations has now made more realistic the testing of a variety of inactivating agents which will provide data about the various structural components and the nature of their relationships within the infectious particle.

The scrapie agent is inactivated by moderate concentrations (1 M) of strong chaotropic ions, such as thiocyanate and trichloroacetate. The inactivation by KSCN suggests that it is a cooperative denaturation process as would be expected with the denaturation of a protein. The surprising feature of these experiments has been the apparent reversibility of this KSCN inactivation. The kinetics and nature of this reversibility suggest that the scrapie agent exists in a form with no less than two critical components which are covalently attached. Based on the experience of the conventional viruses, the assumption must be that scrapie is a nucleic acid with a hydrophobic protein attached.

Despite the existence of these purified preparations and the background body of information available, there has yet to be a demonstration of a scrapie-specific nucleic acid or of a scrapie-specific protein in these preparations. With the exception of an unconfirmed report of DNase sensitivity, there has been no suggestion as to the type of nucleic acid in the scrapie agent. For example, treatment of purified scrapie preparations denatured with KSCN with a variety of enzymatic and chemical agents capable of degrading both DNA and RNA has not led to the loss of scrapie infectivity. These findings leave the virologist with a true enigma. Are we in fact dealing with a truly unconventional agent which lacks any type of nucleic acid as a core of its genetic information? It is unlikely that even these unconventional viruses are so unconventional that they have found a new form of genetic information. However, the dilemma persists, and the current characterization of the agents must be done in light of these seemingly confusing results.

One thing seems certain based on the experience of several laboratory groups. Despite the intellectual appeal to the contrary, it would appear that the unconventional agents do not constitute the first form of animal viroid. Studies from many laboratories have clearly indicated that no infectious nucleic acid can be derived from these agents. And, as might be expected from the evidence just described, phenol extraction of infectious material leads to an immediate inactivation of the infectivity.

Even with the extensive knowledge that has been gained in the past few years, it remains clear why, over the past decade, scrapie has been hypothesized to be a replicating agent without nucleic acid, a self-replicating membrane, a replicating polysaccharide, a basic protein, a viroid, a replicating nucleic acid with a polysaccharide coat, and a hydrophobic nucleoprotein. Based on the evidence cited above, it appears that the best working hypothesis for a model of the scrapie agent is one similar to that of a membrane protein. The model features a small nucleic acid capable of coding for a single polypeptide of approximately 25,000 daltons molecular weight. The evidence for a nucleic acid of this size in this model is derived from the data based on the irradiation target size with ionizing radiation. One copy of the polypeptide is covalently attached to the nucleic acid molecule and hydrogen bonds to it through a series of basic amino acid domains. An unusual feature of this proposed scrapie-specific molecule is its high content of hydrophobic amino acids, leading to a surface with a high degree of hydrophobicity. The resulting structure is one of such extreme net hydrophobicity that it cannot exist in the cytoplasm in a soluble form but must remain associated with lipid-containing structures like the cell membrane. This model incorporates the majority of our current information about the structure of the scrapie agent and is consistent with all the information known about the agent of Creutzfeldt-Jakob disease.

The model presented does not, however, provide a convenient mechanism for replication and production of the scrapie-specific protein. One must assume that at some stage during the replication cycle the scrapie nucleic acid or a transcript of that nucleic acid must finds its way onto a polyribosome in order to make a polypeptide. An alternative, of course, would be a polypeptide produced from host messenger RNA in response to some stimulus triggered by the scrapie nucleic acid. An additional dilemma is that the small size of the scrapie genome precludes its coding for an extremely hydrophobic protein which would require a hydrophilic region as a carrier to get it from the cytoplasm to the membrane, as is the case in many other membrane-maturing viruses.

PATHOGENESIS OF SLOW VIRUS INFECTIONS

The available information on the pathogenesis of infection by the unconventional agents has been derived almost entirely from the study of scrapie. However, the limited available information on kuru and CJD suggests that the scrapie model describes correctly the pathogenesis by each member of the group. The incubation times and temporal sequences of these infections vary widely, dependent on the agent, the species of the host, the route of infection, and the genetic composition of the host. However, the general patterns of the pathogenesis and the relative times of migrations through the body seem to be quite similar.

It is important to understand the problems associated with studies of pathogenesis of these unconventional viruses. The first and foremost difficulty associated with working on these agents is the absence of any type of in vitro assay. The absence of a rapid convenient assay system has hampered not only the study of the physical and biochemical properties of these agents but likewise the study of their pathogenesis and epidemiology.

The following pathogenic pattern was derived by studies of the mouse infected with scrapie by different routes of inoculation. The differences between the routes of infection have a substantial impact on the incubation time and ultimate outcome of the infection. However, the general sequence of the distribution of the agent appears to be similar regardless of the site of initial infection.

Following inoculation of mice with scrapie by either subcutaneous, intraperitoneal, or intracerebral routes, the virus is found entirely in the spleen by 1 week postinfection. The amount of virus associated with the spleen at that time accounts for the material inoculated into the animal and probably does not constitute the results of early virus replication. Following this initial recovery of infectivity of virus in the spleen, there is a decrease or complete loss of infectivity recoverable from any organ of the animal for an additional 2 week period. This "eclipse" period is dependent to some extent upon the route of infection and probably is a reflection of the time it takes to clear the agent from various organs of the body following the initial inoculation. The level of scrapie in the animal begins to show a measurable increase 4 weeks after inoculation. The initial reappearance of the agent is confined to the spleen and lymph nodes. The virus spreads throughout the lymphoid organs of the animal over the next 2 week period, and at this stage of the infection it is quite clear that the level of virus recovered from the animal exceeds the inoculum by manyfold, indicating that multiplication has taken place in these tissues. Despite the fact that virus replication is going on throughout the immune system, there is no notable histopathology or obvious impairment of immune function at this stage of virus infection. The first appearance of any infectivity in the nervous system occurs approximately 3 months after the initial infection of the mice and is found associated with the spinal cord. Shortly thereafter, the agent clearly moves into the entire central nervous system and the levels of virus continue to increase until the terminal stages of the disease. At the time of the first appearance of the virus in the brain it has also become widely distributed throughout the body. It is recoverable from the lymphoid tissues, salivary glands, lungs, and intestine. Despite the high level of virus found widely distributed throughout the animals at this time, they remain totally asymptomatic and do so until approximately 23 weeks after inoculation when some animals begin to show signs of early neurologic disfunction. This stage coincides with the beginning of histopathologic evidence of the disease in the brain. Also at this time the maximum titers of the agent in the brain have been achieved, and the level of virus in the lymphoid tissues has begun to decline. From this point through the terminal stages of infection the titer of the virus in the brain tissue remains virtually unchanged, while the titer in the lymphoid tissues continues to decline.

This mode of pathogenesis of scrapie in the mouse is not unlike that seen with virus infection of several types, for example, lymphocytic choriomeningitis virus. Following the initial infection the virus is almost entirely associated with lymphoid tissue in which it replicates and spreads, moving on to the central nervous system, where it replicates to high titers while at the same time disappearing from the lymphoid tissues. It is clear in the case of scrapie that replication goes on in the brain several weeks prior to the onset of disease. The earliest pathologic change noted during this period is astrocytic hypertrophy and not neuronal disintegration or status spongiosis. There are several striking differences between the experimental infections of the mouse and the goat. In mice, high titers of the scrapie agent are associated with the salivary gland and intestines, while in the goat the nasal mucosa, salivary gland, and intestine contain little or no virus.

Different routes of inoculation which are not so well studied as the subcutaneous route appear to give the same general pattern of infection. There may be, as mentioned earlier, no absolute eclipse period following intracerebral inoculation, but it remains clear that replication of the agent goes on in lymphoid tissue prior to replication in brain tissue, even following intracerebral inoculation.

It is not well understood what cells are associated with replication of the agent in lymphoid tissue. It is clear that no measurable impairment of the immune system occurs in these animals. Therefore, if the lymphocytes are the infected cell type, it clearly does not alter their functional capacity. Congenitally asplenic or splenectomized mice are capable of replicating the virus, although the course of infection may be somewhat prolonged. Likewise, congenitally athymic animals appear to have a normal course of disease. The genetic makeup of the animal appears to play a more profound role upon the incubation time than does the cell biology of the lymphoid system. Investigations of the cell type associated with the spleen infectivity have suggested that it is not associated with the macrophages or other adherent cell populations but is most likely associated with the T or B lymphocytes. The best available evidence suggests that infectivity in the mouse is associated with the B lymphocytes.

It is interesting to note that infectivity is never recovered from acellular body fluids. Extensive investigations of the presence of virus in the saliva and urine have both shown the complete absence of infectivity. Infectious material in the blood has only been associated with the cellular blood components and is never found associated with the serum or plasma alone. These observations are consistent with the apparent pattern of the spread of the agents throughout the infected host. This spread appears to be entirely associated with the migration of lymphoid cells early in infection.

Careful studies on the influence of inoculum size on the outcome of the disease have revealed a strict correspondence between the injected dose and the incubation time in both scrapie-infected mice and hamsters. This relationship is linear between the log of the injected dose and the time of appearance of symptoms or time of death. This occurrence appears to be the result of the exponential replication of the agent. The assumption is that, regardless of the infectious dose, the titer of virus in the brains of the animals reaches the same level before clinical signs of disease occur. There is no direct experimental evidence to support this assumption. However, this dose-time relationship is the basis of the incubation time assay used by some investigators. This assay has been extremely valuable in reducing the number of animals needed for an experiment, as well as in reducing the time needed for the assay.

Limited studies on the pathogenesis of a mouse-adapted strain of CJD following intracerebral inoculation of mice have shown a pattern identical to that just described for scrapie. While virtually nothing is known about the pathogenesis and spread of CJD and kuru in humans, it is clear that the distribution of infectivity in patients at the time of autopsy is consistent with the distribution of infectivity in scrapie- and CJD-infected mice. All patients whose organs have been examined for these viruses have had the highest infectivity associated with brain tissue, and only rarely has virus been recovered from organs outside the central nervous system. The recoveries from peripheral organs have been from spleen, lymph node, liver, and kidney.

The linear relationship between the log of the dose and the incubation time described above for scrapie in the rodent is also true for CJD in the mouse. The exact time of incubation for a given dose and the slope of the dilution curve depends on the strain of mouse being used for the assay. The linear relationship between dose and incubation time does not apply to CJD in the guinea pig. In this host it is not uncommon for animals with a two log difference in inoculum to develop disease simultaneously.

While the information available about the pathogenesis of the unconventional viruses in the animal is sketchy, our knowledge of the mechanism of pathogenesis at the cellular level is even less well developed. While infectivity has been associated with cells held in culture following establishment from an infected animal, there has never been a verified report of the transference of infectivity in cell culture or the existence of a cell line in which every cell was infected with the agent.

There have been several attempts to measure cellular changes associated with scrapie infection in mice. One approach to these studies has been the examination of the rate of DNA synthesis in infected tissue. An increase in the rate of DNA synthesis in the brains of mice infected with scrapie when compared to control mice has been reported. However, the synthesis was mostly in ependymal and glial cells, and autoradiographic studies have shown that this increased synthesis is not associated with areas of pathology. Other workers have observed an increase in the incorporation of DNA precursors into DNA-polysaccharide complexes in both the cytoplasm and the membranes of brains of infected mice. These fractions were also low in infectivity, supporting the notion that all these phenomena are secondary and not related to the direct effects of infection at the cellular level.

At the electron microscopic level, cellular changes specific to virus infection are readily observed. Areas of vacuolation and "swelling" of the cytoplasm have been described in areas of the most intense pathology. These vacuoles appear to begin with focal clearing of the cytoplasm and replacement of normal cellular organelles by a fine granular material. The cleared areas are surrounded by membranes which appear to have ruptured and generated extramembraneous material. These vacuoles were observed in the cytoplasm of neuronal axons, dendrites, and perikarya. This same investigation also described a limited amount of cell fusion, including fusion of astrocytes and neurons. Based on this electron microscopic evidence, Lampert hypothesized that the site of the damage at the cellular level was the plasma membranes, an hypothesis consistent with the biophysical data cited above.

The dilemma in studying the cellular changes associated with virus infection appears to be related to finding the appropriate cell type for in vitro infection and finding a way to measure the effects of the infection process. In the absence of any form of cytopathic change related to the infection, there is at this time no reliable means to measure the presence or replication of the agents within a given cell type. The observations that there seem to be preferential replication or association of the agents with certain differentiated cell types suggests that in vitro cultivation of the agents may be limited only to those cells of the appropriate differentiated type. These differentiated cell types are invariably the most difficult to maintain in culture and this has, no doubt, hindered this work.

The conclusion that one must reach is that the study of the pathogenesis of the spongiform encephalopathies has just begun. It is not at all understood how the agents cause pathology and the characteristic changes at either the organismic or cellular levels. It is clear that these agents lead to profound and devastating changes, most likely by alteration of the ability of differentiated neurons to effect their function. The observations that specific subclasses of neurotransmitter receptor sites are missing or reduced in terminally affected scrapie-infected hamsters supports the notion that the mechanism of cell pathology lies at the membrane level and involves an irreversible change of membrane structure or function in localized areas. These observations have not been confirmed and probably should be reexamined in the near future.

INTERACTIONS WITH THE IMMUNE SYSTEM

The viral spongiform encephalopathies cause extensive pathologic change in the central nervous system without any evidence of inflammation. This feature of the diseases is apparent not only in the natural host but likewise in a wide variety of experimental animals to which they have been adapted. Despite efforts by many laboratories, none have shown any evidence for either humoral or cell-mediated immunity. Attempts to measure humoral antibody have utilized neutralization, antiglobulin-potentiated neutralization, and indirect immunofluorescence on a variety of potential target tissues, as well as the examination of tissues for immune complexes. In addition, sera from experimental animals and patients have been examined for anti-nucleic acid antibody, all without success.

It is quite clear that infected hosts do not have any impairment of their immune responses. Animals infected with scrapie are able to mount an immune response to a variety of immunogenic substances. Likewise, investigations have shown that the mitogenic response of both B and T lymphocytes in infected animals is normal, with the possible exception of a transient depression of murine B-cell mitogenic responses at approximately 30 days postinfection. Other investigations have shown that the cellular immune response in infected animals shows no measurable impairment. These studies compared mixed lymphocyte cultures of splenocytes from scrapie-infected and normal mouse brain-inoculated control animals. In addition to the demonstration that splenocytes from scrapie-infected mice were normal in their mixed lymphocyte response, this same study used this in vitro stimulation system to show that infected spleen cells have no scrapie-specific antigens on their surface. The bulk of the available evidence points to the fact that there is no pathology in the immune system in infected hosts and that the absence of the immune response is simply a reflection of the absence of an immunogenic molecule exposed to the immune system.

Additional evidence that the immune system plays no role in these slow infections is derived from the fact that scrapie-infected mice are unaffected by the state of their immune system. Both thymectomized and nude mice show incubation periods indistinguishable from that in their immunologically competent counterparts. Likewise, immunosuppressed mice show incubation periods identical to that of normal mice.

GENETIC CONTROL OF HOST SUSCEPTIBILITY

The influence of the host genetic background on the susceptibility and the incubation time of scrapie was first studied about 20 years ago in different flocks of naturally infected sheep. Despite the recognition that this disease was transmissible from sheep to sheep and from sheep to goats, the original genetic analysis suggested that the disease was due to "a single autosomal recessive gene." Detailed genetic analysis of scrapie susceptibility in sheep has been complicated by maternal inheritance and by the coinfection of various flocks of sheep with several different strains of scrapie. However, despite its difficulties, genetic analysis of susceptibility to these agents is of great significance because of the unique biochemical and physical properties of the agents coupled with the ill-defined mode of pathogenesis. One would hope that an examination of the genetic alteration of the host would provide additional insight into the mode of transmission of these diseases between different individuals as well as the mode of pathogenesis within a given host. The most extensive studies on the genetic basis of the host interaction with scrapie has been done by Alan Dickinson and his colleagues in Edinburgh, Scotland. Their work has involved an extensive analysis of the genetics of sheep, as well as the detailed analysis of susceptibility and incubation time in mice. Because of the peculiarities of the infection by the unconventional viruses, it is frequently difficult to determine whether an animal is truly resistant to infection or whether in fact the genetic makeup of the host has extended the incubation period to exceed the life span of the animal. In addition to the intrinsic complications of susceptibility to an agent with an unusually long incubation period, there is the additional complication of variations of incubation period within strains of scrapie.

Genetic analysis of the susceptibility of several groups of sheep to either subcutaneous or intracerebral inoculation of the scrapie agent led to the classification of the animals in the flock into short

(197 ± 7 days) and long (917 ± 90 days) incubation classes. When sheep with known short incubation periods were mated with sheep with longer incubation periods, the results suggested a single dominant gene for susceptibility or short incubation time. Following this work, several investigators have established stocks of sheep selected for resistance and now have animals with greater than 90% differences in their susceptibilities to scrapie following subcutaneous inoculation. The available genetic evidence supports the idea of a dominant allele which confers short incubation time and susceptibility to scrapie in sheep. It is clear that sheep with the homozygous short incubation time marker uniformly show incubation periods on the order of 178 to 180 days, whereas animals selected for long incubation periods or resistance to scrapie show a small percentage (approximately 5%) that develop scrapie after incubation periods of more than 1000 days. The presence of those animals which develop disease after long incubation periods has yet to be resolved but complicates the interpretation of the long incubation period genes in sheep.

Following the successful adaption of the scrapie agent to the mouse, Dickinson and his colleagues did careful genetic examination of the susceptibility of mice to scrapie. In their initial screening of different strains of mice, they found that all mice were susceptible to the scrapie agent but that the incubation periods between inbred strains vary between 20 and 40 weeks. When complete genetic analysis was carried out in crosses between short incubation period mice and long incubation period mice, it was clear that the F1 hybrids had intermediate incubation periods while the F2 mice and backcross animals showed segregation of the incubation period time marker, and the segregation pattern was consistent with the idea that there was one gene controlling incubation period in mice and that that gene had two alleles with neither allele being dominant.

More extensive studies in the same laboratory revealed that the situation of the genetic control of the incubation period was more complicated than had been initially thought. In the initial experiments, the C57Bℓ mice, which have a short incubation period, were said to have the "*sinc*" allele and another strain, the VM strain, the long incubation time allele of the same gene. When these mice were challenged with a different strain of scrapie the situation became reversed. Additionally, F1 hybrids between the C57Bℓ and VM mouse exhibited the phenomenon of overdominance. In these F1 animals the incubation period for scrapie was in excess of that for either of the parental strains. Additional genetic crosses showed that the genes controlling the incubation period segregated as a single gene, and the argument has been that the *sinc* gene is also responsible for these effects. Support for the idea that the *sinc* gene controls the same phenomenon was derived from the challenge of F2 and F3 generation mice from a

C57Bℓ times VM cross with both strains of scrapie, the ME7 strain which shows a short incubation period in the *sinc*-carrying animals and 22A strain which shows a short incubation in the non-*sinc*-carrying mouse. In these experiments there was no segregation of the response to each strain of scrapie, suggesting that a single gene or a very closely linked pair of genes was responsible. In addition, the response of a given mouse strain to infection by a defined strain of scrapie could be used to predict the response of that mouse to other strains of scrapie.

The Edinburgh workers have also tried to examine the pathogenesis of the disease in animals carrying different alleles of the *sinc* gene. They have demonstrated that in long incubation period mice the scrapie agent appears in the spleen at a much later time in infection than it does in a more rapid incubation time mouse. Whereas incubation of the scrapie agent ME7 in the spleens of short incubation period mice is almost immediate following inoculation, in the long incubation period mice there is a delay of approximately 4 weeks before the scrapie infectivity begins to appear in the spleen. Following the initial difference in the time of appearance, however, the amount of virus produced in the tissue and the persistence of that virus was approximately the same in both strains of mouse. The appearance of multiplication of the virus in the brain was also delayed in the long incubation period mouse, although the ultimate titer achieved during infection of the mice was the same as it is in the short incubation period mouse.

The ultimate explanation for the activity of the *sinc* gene and the explanation of its role in the pathology and pathogenesis of the disease is not at all clear. Dickinson has hypothesized that the gene may play a role in coding for a heteromeric replication site that could vary in its efficiency of replication depending on the allelic composition of that heteromeric structure. To date there is no firm evidence of what role this gene may be playing in these animals, nor is there any evidence for a specific replication site in cells.

Recent studies of the genetic control of incubation time of mice infected with Creutzfeldt-Jacob disease done in this laboratory have suggested that the picture of genetic control of incubation time is somewhat different than that described for scrapie. An analysis of various mouse strains revealed that the K.Fu. isolate of CJD showed a range in incubation times ranging from 90 to 170 days. By analysis of a series of inbred strains of mice identical except for minor variations in the H-2 haplotypes, as well as looking at a series of mice with identical H-2 haplotypes in different genetic backgrounds, it became clear there were two genes controlling the incubation period. One of these genes is H-2 linked and appears to map at the D end of the H-2 locus. The other gene is a non-H-2 linked gene, the full nature of which has not yet been determined. There has not yet been any physiologic analysis of the influence of these genes on the pathogenesis of the disease. However, it is interesting that cell surface components in the form of the H-2 antigens seem to be intimately involved in the incubation period of this disease.

EPIDEMIOLOGY OF THE SPONGIFORM ENCEPHALOPATHIES

Kuru is a disease confined to the Eastern Highlands of Papua, New Guinea. In that region it is mainly associated with the Fore linguistic group and presents as a subacute degeneration of the CNS. Clinically it is characterized by cerebellar ataxia and a shivering-like tremor (hence, its name: *kuru* means shivering or trembling in the Fore language). The disease progresses to complete motor incapacity and death, usually in less than a year from the onset of the disease. Kuru was first described by Gajudsek and Zigas, who recognized its heterofamilal clustering and limited geographic distribution but were unaware of the infectious etiology of the disease. It was the recognition by Hadlow that the pathology of kuru resembled that of scrapie that led to his suggestion that the two diseases may have a similar etiology. It was this suggestion that led Gajdusek and his colleagues to inoculate the brains of kuru patients into chimpanzees and to watch the progression of these animals over a long period of time. The ultimate success of the transference of this disease to higher primates led to the subsequent transmission of the other human spongiform encephalopathy, Creutzfeldt-Jakob disease, to chimpanzees by the same laboratory.

It is now recognized that kuru was introduced and spread through the Fore population through the practice of ritualistic cannibalism. Because of the cannibalistic ritual done by the Fores, the largest number of cases were associated with women and children. Adult males rarely participated in the ceremony and seldom ate the flesh of deceased victims. This factor led to a 3:1 male to female sex ratio in the Fore tibes.

Unlike kuru, CJD is a sporadic disease showing familial clustering in 10 to 15% of cases. CJD occurs worldwide and has been described in every continent. The worldwide incidence of the disease is unknown, but in those areas where reports of the disease are reliable the annual rate is one to two cases per million population. The mode of transmission of CJD is unknown, and there is no compelling evidence to suggest contagion among humans. There have been reports of iatrogenic transmission of CJD through the surgical transplantation of a cornea and the use of contaminated electrodes in patients being stereotactically studied for focal epilepsy.

Table 2 Host Ranges of the Unconventional Viruses

Scrapie	Nonprimate hosts: gerbil, goat, hamster, mink, mouse, rat, sheep, vole Primate hosts: capuchin, spider, squirrel, cynomologus macaque, rhesus Not susceptible: chimpanzee, cat
Kuru	Nonprimate hosts: ferret, mink Primate hosts: chimpanzee, gibbon, capuchin, marmoset, spider, squirrel, wooly, African green, bonnet, cynomologus macaque, mangabey, rhesus, pig-tailed macaque
Creutzfeldt-Jakob disease	Nonprimate hosts: cat, ferret, goat, guinea pig, hamster, mouse Primate hosts: chimpanzee, capuchin, marmoset, spider, squirrel, wooly, African green, bush-baby, cynomolgus macaque, mangabey, patas, pig-tailed macaque, rhesus, stump-tailed macaque

Both kuru and CJD were originally transmitted to the chimpanzee. In both cases the infected animals showed incubation periods ranging from 11 to 14 months following intracerebral inoculation of case tissue. The diseases have been subsequently transferred to a wide variety of old and new world monkeys, as well as the chimpanzee. However, there still remains no in vitro system for the study of these agents, and the transmission of kuru to small animals has yet to be achieved. Only recently has CJD been transmitted to laboratory animals such as the guinea pig and mouse. Table 2 lists the known host ranges for each of the viral spongiform encephalopathies.

The brains of patients and animals with the spongiform encephalopathies show modest atrophy of the cerebellum and cerebrum with microscopic widespread neuronal degeneration and cell loss with a predominant gliosis. The spongiform change in the grey matter of the cerebral cortex and spinal cord is characteristic of the diseases. This spongiform change appears to represent vacuolar enlargements within the neuronal processes and appears to be of intradendritic origin. Within these swellings there appear to be accumulations of membranous material. Similar vacuoles are also seen in experimental spongiform encephalopathy, including transmitted kuru, CJD, and scrapie. This histopathologic pattern appears to be independent of the host for the experimental disease. The pathology in these diseases is strictly confined to the central nervous system, despite the fact that the agent can be found in various tissues throughout the infected individual.

Although scrapie has been recognized for hundreds of years and has been more intensely studied than any of the other spongiform encephalopathies, the mechanism of its natural transmission remains obscure. The naturally acquired infection appears to transmit from one infected sheep to other sheep and goats. However, such horizontal transmission has never been seen in experimental infections. Scrapie appears to pass from infected ewes to their lambs even if the lambs never suckle from their mother. The early contact between a mother and her lamb seems to be adequate. It is well known that the placenta is infectious, and this alone may constitute an adequate exposure. Older sheep develop the disease only following prolonged contact with diseased animals. Susceptible sheep have developed scrapie by simply occupying pastures previously occupied by scrapied sheep, sometimes several years later.

The neurologic symptoms of scrapie are characterized by progressive ataxia, wasting, apprehension, aggressiveness, and on occasion dementia. Early in the disease tremors of the head and neck are seen, and these progress to become more generalized later in the disease. Fasciculations of superficial skeletal muscles frequently occur and the resulting rubbing and scratching to produce cutaneous irritations are characteristic. The microscopic pathology is confined to the CNS and closely resembles that described above for the human diseases.

Transmissible mink encephalopathy is almost identical to scrapie in both clinical picture and microscopic pathology. The apparent source of the disease was the feeding of scrapied sheep carcasses to ranch mink, and the disease is generally considered to be scrapie. The clinical disease can be easily reproduced by the inoculation of sheep or mouse scrapie into mink.

BIOHAZARDS AND THE AGENTS OF THE SPONGIFORM ENCEPHALOPATHIES

Although the natural transmission of the spongiform encephalopathies is understood only in the case of kuru, there are specific instances of known transmissions of CJD. These transmissions were, in each case, associated with an invasive medical procedure. The requirement for very small doses when introduced directly into the CNS was clearly demonstrated in a Swiss clinic, where a CJD-contaminated stereotactic electrode was the means of inoculation. A silver stereotactic electrode implanted in the brain of a CJD patient, prior to the onset of the symptomatic stage of his disease, was subsequently implanted into the brains of two patients with intractable epilepsy. Between each patient the electrode was sterilized in a conventional manner, employing ethanol and formaldehyde vapor. The two recipients of the electrode each developed CJD following an 18 month incubation period. Use of the electrode was discontinued after the second transmission, although at the time the contamination was not recognized, and the electrode stored in formalin. Two years later the electrode was cut into pieces and implanted into the brain of a chimpanzee which contracted the disease following an 18 month incubation period. The only other well-documented accidental transmission resulted from a corneal transplant from an infected donor. Other suggestive occurrences include the clustering of four cases of CJD in a single English neurosurgery clinic within a 2 year period and the appearance of the disease in a neurosurgeon who also performed frequent autopsies.

In contrast to the occurrences of iatrogenic transmission cited above, there is no evidence of direct person-to-person transmission of either kuru or CJD. A recent study of 1435 CJD cases failed to identify any common risk factor with the possible exception of neurosurgical procedures. This study generated no evidence for a higher risk among health care or animal care personnel.

One of the most serious biosafety problems associated with these agents is the extreme difficulty in achieving their complete inactivation. As cited earlier (see Table 1), these agents are quite resistant to many chemicals and methods normally used to decontaminate or disinfect. Better methods of destroying these agents are clearly needed. One problem associated with the search for better decontaminants is the apparent strain variation between different isolates of scrapie and CJD. Until the questions regarding the best means of decontaminating exposed objects are resolved, it would appear that treatment with sodium hypochlorite (bleach) remains the most effective method short of autoclaving.

SPECULATIONS ON THE NATURAL HISTORY OF THE UNCONVENTIONAL VIRUSES AND THEIR POSSIBLE ROLE IN OTHER DISEASES

In the absence of complete detailed data regarding the biochemical nature of these agents and no clear understanding of their transmission, speculation is all that is available concerning the origins of these agents. A critical question regarding the natural history of these unconventional agents are their identities relative to each other. Is there really only one unconventional virus, scrapie, and has it been transmitted to mink in the form of transmissible mink encephalopathy and to humans in the form of CJD? We can easily envision kuru as beginning as a sporadic case of CJD which then spread rapidly via direct inoculation or injestion through cannibalism. Since the cessation of cannibalism by the Fore, kuru has steadily declined, and no cases have appeared in anyone born since that time.

Another potential source of the virus is a latent infection, horizontally transmitted and activated by some exogenous stimulus, for example, neurosurgery. If this hypothesis is correct, there must be a reflection of the genetics of susceptibility as described earlier for mice and sheep, combined with some unknown mechanism of induction. The possibility that these spongiform encephalopathies are maintained in the population by epidemic spread has no support whatever from the available epidemiologic data.

The role the unconventional viruses may play in other neurologic diseases of unknown etiology is a topic of active investigation. In his 1976 Nobel lecture, Gajdusek speculated on the role of these agents in chronic infections and listed 28 potential diseases, ranging from multiple sclerosis and amyotrophic lateral sclerosis to ulcerative colitis and chronic arthritis. The definition of the role any of these or related unconventional agents play in these chronic infections awaits the development of modern molecular diagnostic procedures.

SUGGESTED READINGS

Gajdusek, D. C. Unconventional viruses and the origin and disappearance of kuru. *Science 197*:943-960, 1977.

Prusiner, S. B., and W. J. Hadlow (Eds.). *Slow Transmissible Diseases of the Nervous System*. Academic, New York, 1980.

chapter 7

RNA Tumor Viruses

RAYMOND V. GILDEN

NCI-Frederick Cancer Research Facility
Frederick, Maryland

ROBERT M. McALLISTER

Children's Hospital of Los Angeles
Los Angeles, California

This chapter will emphasize the type C viruses, a most interesting subgroup of the larger category of RNA tumor viruses. These viruses were initially found to be members of one of the three subfamilies of retroviruses, the first two subfamilies being lentiviruses (visna, maedi, progressive pneumonia viruses) and spumaviruses (foamy viruses). The recommended term for this third subfamily is oncornaviruses or RNA tumor viruses. It originally included type B, type C, and, later, type D viruses, as well as several other retroviruses that have not yet been classified with certainty. In 1958, Berhard proposed that the RNA tumor viruses known at that time be classified on the basis of morphologic criteria into types A, B, and C. Type A particles are of two types; the intracytoplasmic A particles are essentially the core of the type B viruses, whereas the intercisternal A particles in *Mus musculus* appear to be aberrant forms (incomplete, vestigial) related to bona fide type C viruses found in another species of *Mus*. The type D classification, also based on morphologic criteria, was suggested later by Dalton and by Russian workers. Though the common terminology for oncornaviruses is RNA tumor viruses, they have also been called leukosis viruses, or leukemia viruses, and include viruses with sarcomagenic potential. These retroviruses share the common property of multiplication via a DNA intermediate which then integrates into the host cell genome. Members exist as conventional horizontally transmitted viruses and as inherited genetic elements generally present in relatively high copy numbers. These viruses are evidently capable of multiple random integration events. The pathogenic members are associated with a variety of diseases, including malignancies, neurologic disorders, and arthritis in several infrahuman species, but despite a number of claims to date, no convincing isolates have been made from humans.* Given the infrequent isolation from several primate species, this situation may change at any time.

The type C viruses have received the greater share of attention in recent years, and the current evidence indicates that some members of the subfamily acquire malignant potential by the acquisition of cellular transforming genes. The retroviruses have attracted a great deal of interest from various disciplines, and much speculation has centered on the possible roles of inherited viral and viral-related genes in normal development. As an alternate view, we propose that these genetic elements represent one stage in the life cycle of a highly successful parasite, a viewpoint easily understandable based on the recent information that the integrated provirus has a structure remarkably similar to that of certain bacterial transposons. This chapter reviews some of the extensive studies carried out through 1979 by laboratories around the world to characterize these viruses and to determine their role in cell transformation in vitro and in cancer causation in vivo. It also discusses some of the interesting concepts and speculations generated by these viruses.

HISTORICAL BACKGROUND

In 1908, Ellerman and Bang in Copenhagen discovered that leukemia in chickens was transmissible with cell-free extracts containing what was subsequently identified as a type C leukosis virus. In 1910 to 1911, Peyton Rous showed that naturally occurring sarcomas in chickens were experimentally transmissible by what eventually became known as the Rous sarcoma virus, the prototype type C sarcoma or transforming virus. Interestingly, in Rous' experiments, as well as the similar studies of Fujinami and Inamoto in Japan, the sarcoma-inducing viruses were isolated only after the sarcoma tissues had been serially transplanted in chickens. This observation has bearing on the origin of not only type C sarcoma viruses, but also on nontransforming type C leukosis viruses. After the initial era (1908 to 1935) encompassing the early studies of avian leukosis and sarcoma viruses, the second era of evolving information regarding type C viruses and related viruses (1936 to 1950) saw the discovery by Bittner of the first type B

---

*Recently, R. C. Gallo and associates at the National Cancer Institute have provided compelling evidence that a specific subtype of human leukemia is associated with a horizontally transmitted retrovirus designated human T cell leukemia virus (HTLV). Workers in Japan have now reported similar findings.

virus, the mammary tumor virus of laboratory mice. The third era (1951 to 1970) was initiated with the discovery by Gross of the murine leukemia virus and the demonstration of its vertical transmission. Thereafter, a number of other mammalian type C viruses were isolated, including mouse leukemia (Friend, Moloney, Rauscher strains, and others) and mouse sarcoma (Moloney, Harvey, and Kirsten), cat leukemia (three serotypes) and cat sarcoma (several distinctive isolates), and hamster and rat viruses. In addition, type D viruses of the guinea pig and the as yet unclassified leukemia virus of the domestic cow were discovered. Also discovered during this period was the prototype type D virus, Mason-Pfizer virus, from the rhesus monkey. Along with these mammalian viruses, a retrovirus was also isolated from Russell's viper. This era witnessed an increased use of tissue culture techniques allowing for the isolation and cultivation of type C viruses in vitro. It was observed that type C viruses did not cause cell death as do the usual cytolytic viruses, and also, most strikingly, that the sarcoma viruses caused cell transformation in tissue culture. A major discovery in 1970, that of the RNA-dependent DNA polymerase enzyme (RDDP) by Temin and Baltimore, had great impact on the theoretical development of molecular biology and also provided many new approaches for studies of the RNA tumor viruses. In the fourth era (1971 to the present), a number of additional type C viruses were isolated and identified, including woolly monkey sarcoma, gibbon ape leukemia, RD-114 from the domestic cat, pig, and several related viruses from the rhesus monkey, the stumptail macaque, and the colobus monkey. A remarkable event in this era was the creation by laboratory manipulation of a stable rat sarcoma virus by Rasheed and colleagues. In addition, other type B viruses in *Mus* species and type D viruses of the Langur and squirrel monkey were isolated.

CHARACTERISTICS

The type C viruses can be distinguished from both type B and type D viruses on the basis of their pattern of virion assembly which terminates by budding through the plasma membrane (Fig. 1). There are also differences in the divalent cation preference of their RNA-dependent DNA polymerase. In addition, type C virus structural proteins can be distinguished immunochemically from those of type B and type D viruses, although there is an unusual cross-reactivity among the envelope glycoproteins of primate type C and the type D viruses. Table 1 summarizes the characteristics of type C viruses.

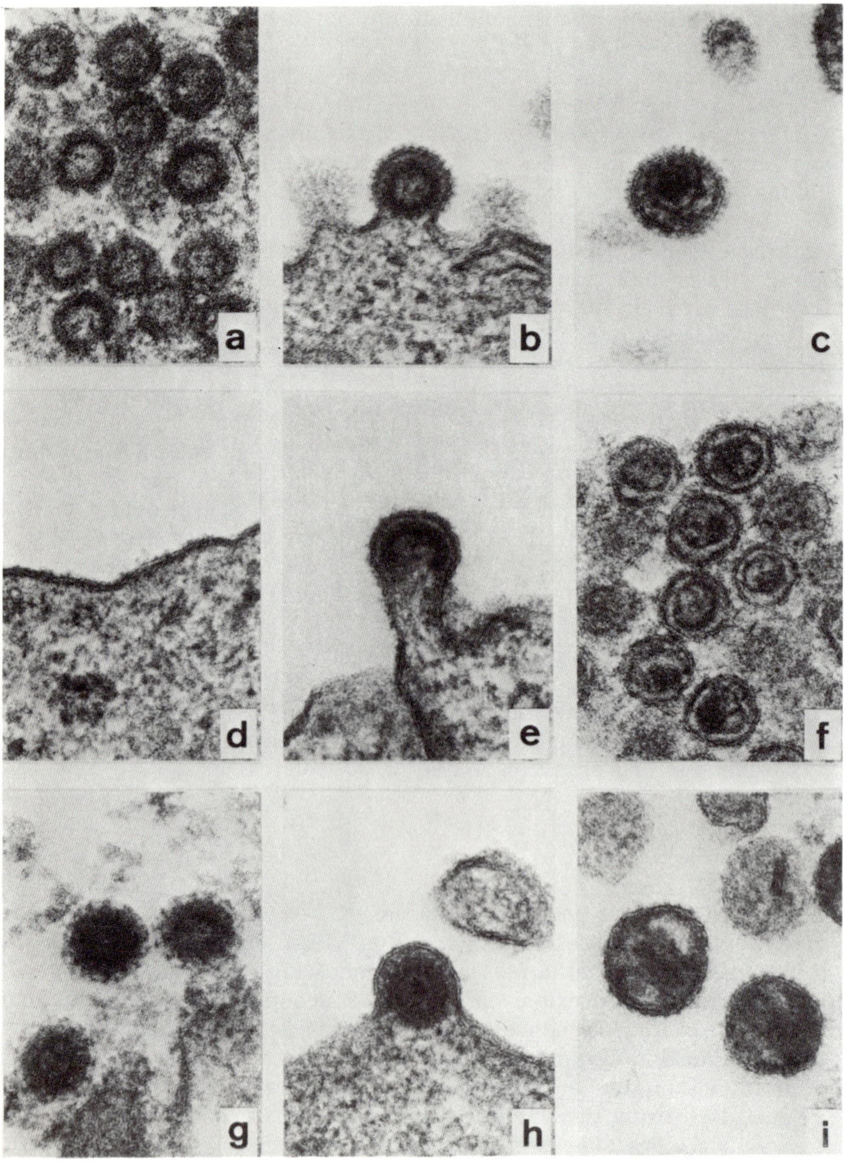
a
b
c
d
e
f
g
h
i

Morphology

The type C virions are spherical, approximately 100 nm in diameter, with an outer envelope and small surface spikes and knobs, an inner membrane, and a central nucleoid (Fig. 1). In contrast, type B viruses have larger and more distinct surface projections and an asymetrically located nucleoid. Type D particles also have an eccentric nucleoid, but are larger than both type B and C particles and have no surface spikes.

Figure 1 Comparative morphology of type B, C, and D retroviruses. (a) cluster of intracytoplasmic type A particles (representing immature precursor of budding type B virus) with doughnut-shaped nucleoid and electron-lucent center; (b) immature type B virus with doughnut-shaped nucleoid and electron-lucent center budding from cytoplasmic membrane; (c) mature extracellular type B virus with eccentric electron-dense nucleoid and characteristic surface spikes; (d) lack of morphologically recognizable intracytoplasmic form for type C virus; (e) immature type C virus with crescent-shaped nucleoid budding from cytoplasmic membrane; (f) mature extracellular type C virus with centrally located condensed nucleoid; (g) intracytoplasmic type A particle with electron-dense doughnut-shaped nucleoid and projections on the outer perimeter of the nucleoid (representing immature intracytoplasmic precursor form of type D virus); (h) immature type D virus with doughnut-shaped nucleoid with electron-lucent center budding from cytoplasmic membrane; (i) mature extracellular type D virus with condensed bar-shaped nucleoid. All micrographs are approximately x87,000. Thin sections were double-stained with uranyl acetate and lead citrate. Type B viruses were MMTV from C3H/Cgrl mammary adenocarcinoma cell cultures. Type C viruses were Theilen strain of FeLV cultivated in FL74 feline lymphocytes. Type D viruses were MPMV grown in A204 cultures. (From D. Fine and G. Schochetman, Type D primate retroviruses: A review. *Cancer Res.* *38*:3123-3139, 1978. Photograph courtesy of M. Gonda, Frederick Cancer Research Center.)

Table 1 Characteristics of Type C Viruses

Capacity to induce leukemia, lymphoma, and sarcoma; some viruses can induce other diseases or be nonpathogenic

Typical morphology with budding forms

Buoyant density 1.15-1.18 g/ml

Single-stranded RNA genome consisting of an inverted dimer with two monomers joined near their 5'-ends; $0.6\text{-}1.2 \times 10^7$ daltons

Viral genes	Gene Products	
gag[a]	p15	Type, group, and interspecies determinants
	p12	Major virion phosphoprotein; type and group determinants
	p30	Major internal antigen; group and interspecies determinants
	p10	Ribonucleoprotein; group and interspecies determinants
pol	RDDP	RNA-dependent DNA polymerase activity, RNAase activity; group and interspecies determinants
env[a]	gp70	Viral envelope glycoprotein; type, group, and interspecies determinants
	p15(E)	Membrane protein, type, group, and interspecies determinants
src	pp21-100	Present only in transforming (sarcoma) type C viruses; phosphoproteins with protein kinase activity

Capacity to rescue sarcoma viruses; nontransforming "helper" viruses

Capacity to induce syncytia in vitro

Capacity to induce cell transformation in vitro

Proviral DNA: endogenous proviruses present in host cell genome in multiple copies; can be infectious

[a]Size of individual polypeptides may vary in different isolates.

Physicochemical Properties

The type C particles have a density of about 1.15 to 1.18 g/cm and a molecular weight of approximately 5×10^8 daltons. Although relatively resistant to ultraviolet (UV) light, their infectivity is abolished by lipid solvents and their half-life at 37°C is 3 to 6 h.

Chemical Composition

RNA

The type C particles have a single-stranded linear RNA genome which sediments at 50 to 70S depending on the particular isolate, corresponding to estimated molecular weights of approximately 0.5 to 1×10^7 daltons. The genomic RNA consists of an inverted dimer structure with two identical monomers ranging from 28 to 35S in sedimentation velocity, joined near their 5'-ends and with distinct secondary features, including loops and hairpins. Each monomer subunit in viruses with a 35S genome consists of approximately 10 kilobases (kb). In leukemia or nontransforming helper viruses, the genome contains three genes, *gag* (group-specific antigens), *pol* (polymerase), and *env* (envelope). In sarcoma viruses, a fourth gene, designated *src*, responsible for the unique biologic activity of these viruses, is present. The genomic RNA accounts for approximately 2% of the virion weight. It is capable of acting as mRNA, but it is not infectious.

Other species of RNA can sometimes be found in virus particles, 4, 5, and 7S, as well as ribosomal 18 or 28S species. One species of tRNA is hydrogen bonded to the genomic RNA and acts as primer for viral RNA-dependent DNA synthesis in vitro. This RNA species has been identified as tRNA trp for chicken type C viruses, tRNA pro for mouse viruses, and tRNA gly for RD-114 and baboon viruses.

Proteins

The type C virion is approximately 60% protein. The RNA genome can code for a translational product of approximately 300,000 daltons. Studies of virus-coded proteins have now identified a sufficient number of individual components to account for this coding capacity. These include structural components coded for by the *gag* and *env* genes, RNA-dependent DNA polymerase enzyme (RDDP) coded for by the *pol* gene, and in sarcoma viruses, a protein associated with malignant transformation coded for by the *src* gene. The six to eight structural polypeptides are designated as gp (glycoprotein), pp (phosphoprotein), or p (protein) and are followed by the first two digits of their molecular weight expressed in thousands. The immediate translational products (Fig. 2) of the three structural genes are polyproteins designated Pr (precursor) for nonglycosylated products or gPr for glycoprotein

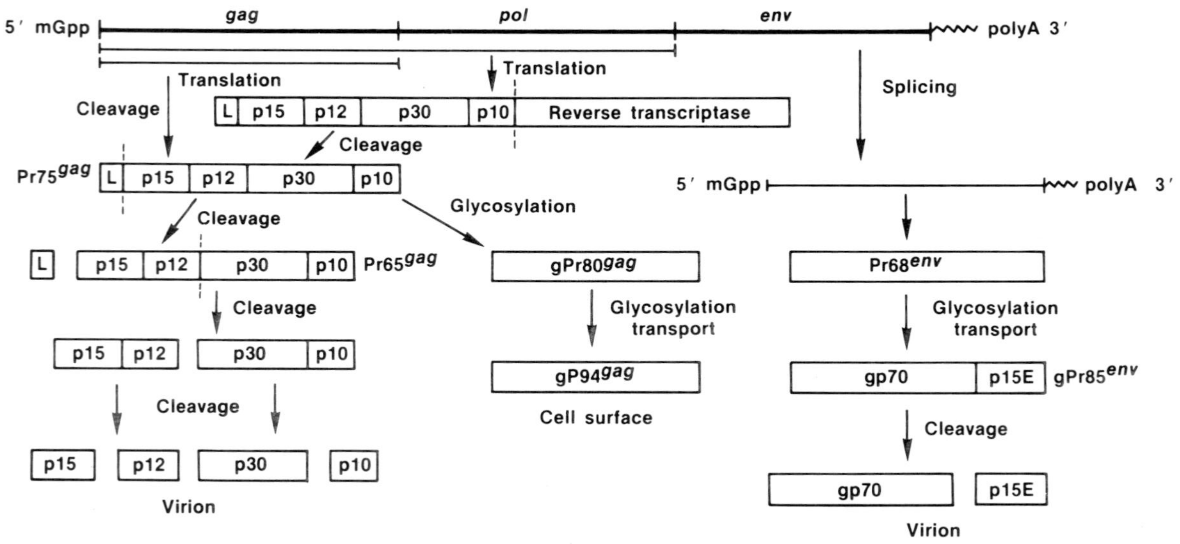

Figure 2 Translational products of a typical type C virus. *Note*: The reverse transcriptase appears as a 70,000 dalton protein in mature virions. (Scheme courtesy of S. Oroszlan, NCI-Frederick Cancer Research Facility.)

precursors. Thus, for one strain of murine leukemia virus (Rauscher), the *gag* gene-coded precursor is a 75,000 dalton product (Pr75) containing a leader peptide (lost during maturation) and, in amino to carboxyl terminal order (5' to 3' in the RNA genome), the polypeptides designated as p15, pp12, p30, and the nucleoprotein p10. These latter polypeptides are found in the mature virion and are generated by specific proteolytic cleavage. The generation of mature virion polypeptides is dependent on phosphorylation, while an alternate pathway involving glycosylation generates a cell surface *gag* polyprotein of 85,000 daltons. This latter product is not incorporated into virions, and the ratio of it to virion *gag* products may vary greatly, depending on virus and virus-cell interaction. The *pol* product is evidently formed by read-through of the *gag-pol* mRNA, overcoming a stop codon at the *gag* terminus. This read-through occurs at approximately 4% of the frequency of termination and results in synthesis of a 180,000 to 200,000 dalton product containing *gag* and *pol* determinants. Subsequent cleavage generates a 70,000 dalton enzyme for mammalian type C viruses; however, larger intracellular forms have been isolated. Differences in Mr are found among the retroviruses, and the avian enzyme is composed of two subunits. The Rauscher *env* primary product, Pr68, is initially glycosylated to give gPr85, which is then cleaved to give the virion products gp70 (surface projections) and p15(E), an integral membrane protein. While the above details the precursor processing of one typical type C virus, there are variations in virion polypeptides, presumably based on shifts in cleavage sites in the precursor molecules. For example, for the primate type C viruses, the NH_2-terminal polypeptide is p12, while the phosphoprotein is 15,000 daltons; however, despite size variations, the corresponding polypeptides remain in the same functional order. Avian type C viruses have some major differences; the *env* products are gp85 and gp37 and the carboxyl terminal *gag* polypeptide, p15, possesses proteolytic activity. Aside from the capacity to use RNA as a template to synthesize complementary DNA, the RDDP of both chicken and mammalian type C viruses also exhibits a ribonuclease H activity which can degrade the RNA strands of RNA-DNA hybrids. Several other enzymes, probably of cellular origin, have been associated with virus particles, such as endonuclease and DNA ligase. The *src* gene apparently codes for only one protein. In chicken type C viruses, this protein has a molecular weight of 60,000 daltons, is phosphorylated (designated pp60 *src*), displays the enzymatic activity of a protein kinase, and is located in the cell membrane. In the mammalian type C transforming viruses, a number of *src* gene products have been reported, ranging in size from 21,000 to 100,000 Mr. They are phosphorylated and many are reported to have kinase activities, and at least one rat sarcoma virus gene product is located on the inner surface of the cell membrane.

Lipids

Lipids account for approximately 35% of the weight of the virion and are derived from the host cell.

Carbohydrate

Type C virions contain approximately 3% carbohydrate, which is synthesized by host cell enzymes.

Replication of Type C Viruses

As noted earlier, type C viruses replicate by means of a DNA intermediate. Following attachment and penetration of a virus particle, the genomic RNA is transcribed into DNA by the virion RDDP. The transcript is integrated into the nuclear DNA of the host via a circular intermediate. Progeny RNA is transcribed from the integrated proviral DNA template by cellular RNA polymerase. The viral mRNA is polycistronic. Primary translational products are polyproteins, as described previously, which are cleaved into the functional polypeptides.

Recombination

Several types of recombinational events are associated with type C viruses which contribute to the pathogenicity of the virus. First, a recombination between the type C virus genes and cellular *sarc* genes leads to the formation of sarcoma viruses of the chicken and of four mammalian species (mouse, cat, rat, and woolly monkey). In the Historical Background section, it was noted that Rous discovered the chicken sarcoma virus after the transplantation of sarcoma tissues into syngeneic chickens. This was the initial observation of this type of recombination. Recent studies of the chicken system have indicated that the type C viral genes (*gag*, *pol*, and *env*) are located on macrochromosomes, while the *sarc* genes are located on the microchromosomes. Most avian and all mammalian sarcoma viruses are replication-defective, transforming viruses. They are competent for transformation (by the expression of the *src* gene), but require type C viruses for their replication.

In chicken sarcoma viruses, the cellular *sarc* gene has been inserted at the 3'-end of the helper virus *env* gene. In contrast, in the majority of mammalian sarcoma viruses, it appears that there is a preferential site for *sarc* gene recombination toward the 5'-end of the helper virus genome in the *gag* gene.

The history of the rat sarcoma viruses appears to be unique among mammalian sarcoma viruses. Three isolates have been reported (Harvey, Kirsten, and Rasheed); two resulting from virus inoculations into mice and rats and the third from cell culture studies (the Rasheed isolate). The viruses are of interest because, although they were derived from different strains of rats, they code for an immunologically related *src* protein (p21 in Harvey and Kirsten viruses and p29 in Rasheed virus). However, while the three rat sarcoma viruses appear to contain a related *src* gene, the precise genetic structure of the viruses and their *src* genes differ. Harvey and Kirsten viruses were derived by passage of uncloned mouse leukemia viruses through mice, then rats, and, finally, through mice. These transforming viruses are recombinants between a mouse type C virus and an endogenous replication-defective virus-like (30S) RNA sequence in rats. These rat 30S RNA sequences have many of the properties of viral sequences; they are approximately 6000 to 7000 nucleotides in length, they contain a poly (A) stretch at the 3'-end of the genome, and they can be packaged into virus particles by type C helper virus proteins. They are also inducible by halogenated pyrimidines and protein synthesis inhibitors. Harvey and Kirsten sarcoma viruses are composed mainly of these 30S sequences. The first 100 to 150 nucleotides at the 5'-end and the last 1000 nucleotides at the 3'-end are of mouse origin, while the remainder of the genomes are of rat origin. These viruses clearly represent interspecies recombinants. The rat cellular *sarc* gene, however, must have been inserted into the rat 30S genome either before or after recombination with the mouse helper sequences, since most normally expressed 30S sequences do not contain the specific *src* insert nor do they transform cells. Rasheed rat sarcoma virus was derived in a different manner and has a different genetic structure from the Harvey and Kirsten viruses. The virus was isolated by cocultivation of a Sprague-Dawley spontaneously transformed rat cell line (SD-1) that released high levels of endogenous rat type C virus with a 4NQ chemically transformed Fischer rat cell line that had been passaged through a rat. Both the SD-1 and 4NQ cells express rat 30S sequences, but neither expresses the rat *sarc* gene product p29. Neither cell, by itself, produced a transforming virus. The cocultivation of these cells, however, results in the formation of a stable transforming virus. This process has been repeated many times and, thus, is a reproducible phenomenon.

In contrast to the Harvey and Kirsten sarcoma viruses that are composed largely of the rat 30S genome, the Rasheed sarcoma virus does not contain 30S sequences. Instead, the genome consists almost entirely of the rat helper virus genome and, therefore, is similar to other avian and mammalian sarcoma viruses. The *src*-derived p29 product thus represents a fusion protein containing a portion of the amino terminal *gag* protein and p21 *src*. These observations offer a unique and potentially rewarding model system to study the origin and function of mammalian sarcoma viruses and sarcoma virus gene products. By the use of current genetic and biochemical technology, it should be possible to gain an understanding of the process by which mammalian retroviruses act to induce malignancy.

A second type of recombination has been observed between the genomes of different host range classes of murine type C viruses (ecotropic, xenotropic, and amphotropic). Ecotropic viruses grow in cells of the species of origin, e.g., mouse viruses grow in mouse cells, xenotropic viruses grow in cells of foreign species, and amphotropic viruses grow in cells of both host and foreign species. Theoretically, both intercistronic (between the three sets of genes, *gag*, *pol*, and *env*, of the different virus classes) and intracistronic (between *env* genes only of different virus classes) recombination can occur, and evidence for each has been reported. Most notably, the MCF viruses isolated from thymic lymphomas of AKR mice have *env* gene products which have peptides common to both ecotropic and xenotropic virus peptides. Also, they have a dual host range for both mouse cells (ecotropic) and cells of other species (xenotropic). It has been shown that the MCF viruses have recombinant *env* genes with the 5'-end derived from an ecotropic parental virus and varying lengths of the 3'-end derived from a xenotropic virus. This observation seems to be of importance because the increased leukemogenicity of virus obtained from AKR thymus at 6 months of age (MCF virus) compared to virus obtained from fetal or young postnatal AKR mice (ecotropic virus) could be explained by this recombinational event occurring in the thymus of young mice. However, apparently some MCF-type isolates are not leukemogenic.

Capacity to Rescue

It has been recognized for many years that an important property of helper type C viruses was their capacity to rescue defective sarcoma (*sarc*) genes from cells experimentally, but not productively, infected with sarcoma viruses. As described earlier, more recently it has been recognized that helper viruses can also recombine with and rescue endogenous *sarc* genes in certain avian and mammalian species. Of interest, types B and D viruses generally do not have the capacity to rescue such defective sarcoma genes from nonproducer cells; however, the generation of a pseudotype containing type B envelope genes from the mouse mammary tumor virus and the Kirsten *src* gene has been reported. Also noted earlier is the recent discovery of novel species of 30S RNA in cells of the rat and the mouse. These molecules have properties of retroviral sequences, but are distinct from those of rat or mouse type C viruses. They can be efficiently rescued by type C, but not type D, viruses. They seem to represent a new class of defective type C viruses in the cells of rats or mice. Ongoing studies will provide additional information about these molecules and their relation to other retroviruses.

Capacity to Induce Syncytia

After the original observation by Klement that ecotropic mouse type C viruses could induce syncytia in monolayer cultures of rat cells, a number of other type C viruses, such as RD-114 and baboon virus, were noted to have the capacity to alter the membranes of the cells they infect. RD-114 virus is also capable of causing fusion of human lymphoblastoid cells growing in suspension.

Capacity to Transform Cells

Manaker and Groupe were the first to observe focal changes induced by Rous sarcoma virus (RSV) in monolayer cultures of chicken cells. This capacity to transform cultured cells has now been associated with both avian and mammalian sarcoma viruses. Transformation is thought to be equivalent to the process by which normal cells become malignant in vivo. Early findings by Rubin and Temin revealed that some RSV-transformed cells continuously released virus and multiplied indefinitely. Infection of a cell by RSV, a RNA-containing virus, had induced a genetically stable change in the host cell. This curious observation eventually led Temin and Baltimore to the discovery of the RNA-dependent DNA polymerase. Intensive research by many laboratories of temperature-sensitive RSV mutants has established the theory that cell transformation is caused by the expression of the *src* gene of the virus.

Host Range

The envelope gp70 glycoprotein, as well as the genetically determined receptors of the host cell, determine the host range of type C viruses. In studies of chicken type C viruses, it was recognized early that the genotype of the chicken cell was important in determining whether the cell would be infected by various subgroups of chicken type C viruses. In studies of mammalian type C viruses, it was noted that murine viruses could be classified according to their capacity to infect mouse cells and/or cells of other species. Endogenous viruses of other species have been found to be xenotropic, i.e., RD-114 virus and baboon virus do not efficiently infect cat or baboon cells, respectively. The presence or absence of the appropriate receptors for different viruses is an important host range determinant for mammalian viruses; thus, mouse xenotropic viruses appear to use a different family of receptors than do mouse ecotropic viruses.

Antigenicity

The translational products of type C viruses are summarized in Table 1 and Figure 2. Immunologic characterization of the gp70s of mammalian type C viruses have shown the presence of type, group, and interspecies antigenic determinants on the glycoprotein molecule. Gp70s have been most useful in discriminating between the closely related type C isolates (type specific) and less useful in identifying the species of origin (group specific) or the mammalian viruses (interspecies specific). Gp70 glycoproteins represent the major target of neutralizing antibody. The p15(E) polypeptide in the viral envelope is distinct from the *gag* gene-coded p15 and possesses type, group, and interspecies antigenic determinants, as now revealed by monoclonal antibodies. The p30 antigen is the major group-specific virion antigen. Immunologic assays studying such determinants on p30 antigens provided one of the first methods for discriminating between type C virus isolates of different species and identifying the species of origin of candidate human type C viruses. It was noted early that p30 proteins also contained interspecies, designated gs-3, determinants that helped to identify isolates as mammalian in origin. P30 proteins possess little type-specific reactivity. The primary structure of the p30 proteins of mammalian viruses has been studied in detail. Partial amino acid sequences have been determined for viral isolates from the cat, baboon, mouse, rat, gibbon ape, woolly monkey, and rhesus monkey.

The p15 of the mouse, rat, hamster, and cat (FeLV) is analogous to the p12 of the RD-114 and baboon viruses based on sharing of interspecies antigenic determinants. The pp12 of the mouse and the analogous protein pp15 in RD-114 and baboon virus bind specifically to the RNA isolated from the same virus, but not to the RNAs of heterologous type C viruses. Immunologic analysis of these phosphoproteins have demonstrated a degree of type specificity. Application of competition radioimmunoassays to diverse type C viral isolates has been of value in discriminating and identifying closely related viruses. The p10s isolated from murine viruses possess strong group-specific antigenic determinants, and interspecies antigenic determinants have also been detected. The p10 competition radioimmunoassays for tissue extracts may provide a useful tool in the search for viral gene expression in tissues of mammalian species from which infectious type C viruses have not been isolated.

The RDDP also provides useful immunologic data for identification and characterization of type C viruses of diverse origin. Antiserum against the enzyme of a given mammalian virus most strongly inhibits the polymerase activity of the homologous enzyme and, to a lesser degree, the enzymes of mammalian viruses, but not the enzymes of avian type C or mammalian type B or D viruses. The mammalian RDDP proteins, therefore, contain both interspecies- and group-specific antigenic determinants.

MOLECULAR BIOLOGY

Virion RNA

The characteristics of this RNA were described earlier. Figure 3 illustrates the structure of the RD-114 genome RNA. The dimers consist of two monomer subunits which appear in the electron microscope to be identical in length and in secondary structure. The monomer subunits have molecular lengths of approximately 5 to 10 kb

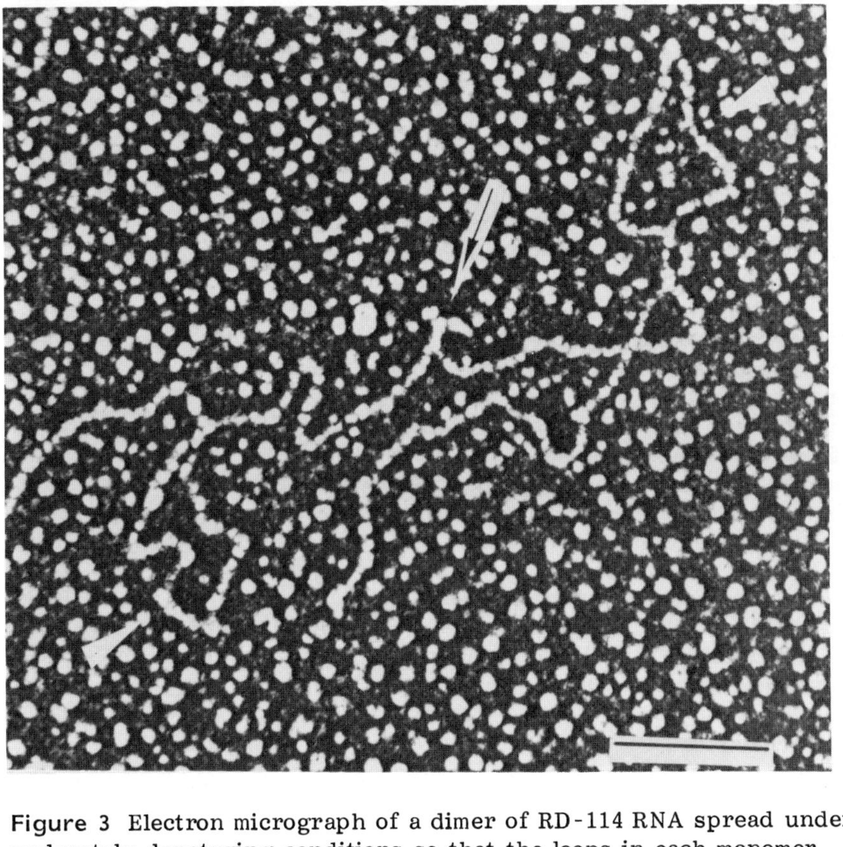

Figure 3 Electron micrograph of a dimer of RD-114 RNA spread under moderately denaturing conditions so that the loops in each monomer arm as well as the dimer linkage structure are preserved. The dimer linkage structure and the two loops are indicated by arrows. (From R. M. McAllister, M. B. Gardner, M. O. Nicolson, and R. V. Gilden, RD-114 virus: Characterization and identification. In *Progress in Experimental Tumor Research*, Vol. 22, pp. 196-215, F. Homburger, Ed., S. Karger, Basel, Switzerland, 1978.)

(1 kb equals 1000 nucleotides or nucleotide pairs). Defective transforming viruses have genome lengths in the 5 to 6 kb range, showing loss of structural gene regions of the parental helper virus. The two monomers are joined close to one end in a secondary structural feature which is called the dimer linkage. Each monomer has a 3' poly (A) end. Poly (A) mapping shows that the 3'-ends are on the outside termini of the primer; therefore, the dimer linkage is close to the 5'-ends of the monomers (Fig. 4). An additional secondary feature, which may be observed in some mammalian retroviral RNAs (Fig. 3), is a loop close to the center of each monomer unit. The viral RNA dimers from various viruses differ in length and in position of the secondary features, such as the loops and hairpins. They also differ in the stability of the dimer linkage under dissociation conditions. Endogenous xenotropic viruses, such as RD-114, baboon, and NZB, have relatively stable dimer linkages; other ecotropic or amphotropic viruses are more easily dissociated. Biochemical evidence, based mainly on oligonucleotide fingerprint analysis, indicates that the two monomer constituents of the dimers are identical in sequence. The dimer linkage is presumably based on base pairing.

The 28 to 35S RNA monomers have termini typical for messenger RNAs. The 3'-ends consist of approximately 200 adenylic acid residues, and the 5'-ends are characterized by a capped structure (presumably providing stability for messenger functions), in which a 7-methyl guanosine links 5' to 5' via a triphosphate to a second 2'-O-methylated nucleotide. This is represented as $M^7G^{5'}$ ppp 5 NmpNp. The 28 to 35S virion RNA can act as messenger both in vivo and in vitro and, therefore, is in the plus polarity. Several lines of evidence support the hypothesis that individual 28 to 35S RNA molecules contain a complete set of genes. Oligonucleotide analysis indicates that the genome of RNA tumor viruses has a unique sequence complexity of 3 to 4 $\times 10^6$ daltons corresponding to a 28 to 35S molecule of single-stranded RNA. The primer tRNA, described above, binds tightly to the homologous 28 to 35S RNA. The binding site is located close to the 5'-end extending from residues 101 to 135 in different isolates.

DNA Provirus

Viral DNA is first synthesized in the cytoplasm approximately 3 h after infection of the cell. At 6 to 9 h, the new "preprovirus" is moved into the nucleus. There it becomes integrated into high molecular weight cellular DNA. Unintegrated viral DNA occurs in the form of linear and open circular molecules and as closed circular double strands. Closed circular supercoiled viral DNA has a size corresponding to the 28 to 35S unit genome of the virus. Both closed circular supercoiled and nonsupercoiled DNA are infectious in transfection experiments. The bulk of infectivity is found in a discrete homogeneous fraction of

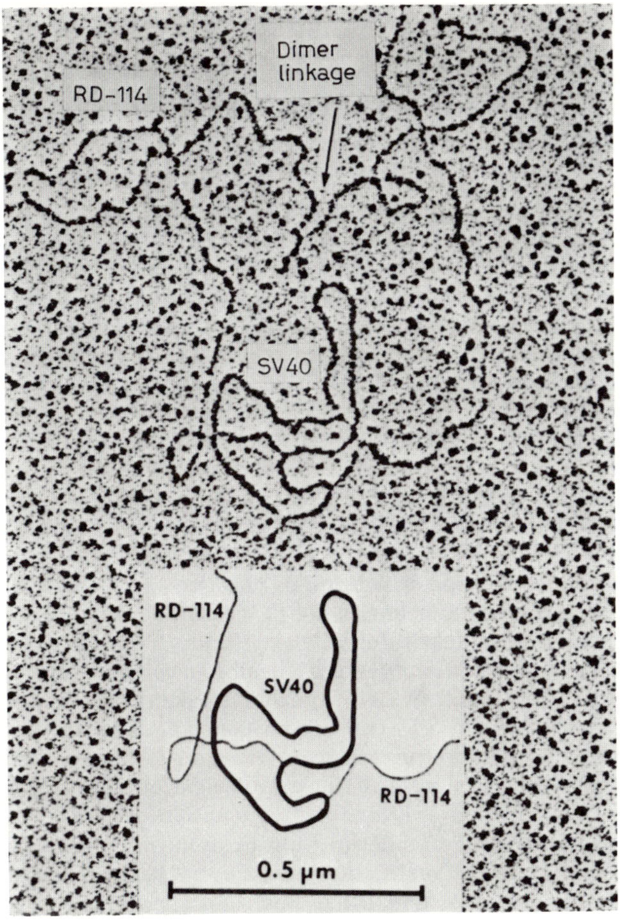

Figure 4 Electron micrograph of an RD-114 dimer. The micrograph shows that the poly (A) segments present on each monomer component of the dimer are at the outside ends. Therefore, the dimer linkage structure is close to the 5'-ends of the monomers. The poly (A) segments can be identified because they are hybridized to poly (dT) tails that have been enzymatically attached to an SV40 duplex circular DNA. The inset sketch shows the detailed topology where the RD-114 ends join the SV40 DNA. (From R. M. McAllister, M. B. Gardner, M. O. Nicolson, and R. V. Gilden, RD-114 virus: Characterization and identification. In *Progress in Experimental Tumor Research*, Vol. 22, pp. 196-215, F. Homburger, Ed., S. Karger, Basel, Switzerland, 1978.)

linear DNA. The size of this component corresponds to one 28 to 35S unit genome. Initially, the mouse sarcoma virus linear DNA was studied with restriction endonucleases and the Southern blot technique and a map of genome fragments was obtained. These techniques have now been applied to the unintegrated, as well as the integrated, proviruses of other avian and mammalian type C viruses.

The 5'- and 3'-Termini of Proviral DNA

The prevalent DNA synthesized from the virion RNA-dependent DNA polymerase represents the 5'-end of the genome extending 200 nucleotides or less from the terminus (strong stop cDNA). Even under conditions which yield some DNA transcripts of full genome length, most of the DNA molecules are small and of discrete length (101 nucleotides in avian sarcoma and 135 nucleotides in murine leukemia virus). They are initiated with the primer tRNA and constitute the sequence from the primer to the 5'-end of the genome. These discrete homogeneous DNA fragments from the 5' end of several viruses have been isolated, purified, and sequenced by the method of Maxam and Gilbert. The sequence of the 5'-terminus has several features which may shed light on important properties and functions of the genome. The first 16-79 nucleotides immediately adjacent to the 3'-terminal poly (A) are a direct repeat of the 16-79 nucleotides at the 5'-end. This terminal redundancy of the viral genome suggests a possible mechanism for reverse transcription. Initiation of DNA synthesis takes place at the tRNA primer and proceeds to the 5'-end of the genome. Transcription of the remaining genomic RNA requires circularization facilitated by the terminal redundancy. In the course of the circularization, the redundant sequences are transcribed only once. Sequence information is now available for several retroviruses following integration into the host genome. Coupled with detailed restriction enzyme analyses, a generalized picture of the integrated provirus has been obtained (Fig. 5). The important features are long terminal repeated sequences (LTR) at either end of the viral genome, ranging from several hundred to almost 1 kb in length. Note that the replication process has resulted in sequences from the 3'-end of the genome being located at the 5'-end of the provirus and the 5'-terminal sequences are also located at the 3'-end. Within the LTR there are sequences resembling known promotor, ribosomal binding, and polyadenylation signals. Other evidence, using cloned subgenomic fragments, indicates that, as predicted, the LTR is critical for establishing or enhancing the biologic activity of coding segments. Note that each LTR contains an inverted repeat, as does the whole genome; thus, this structure resembles certain bacterial insertion sequences, and the entire provirus resembles a transposable element. Integration events evidently occur randomly in host cell DNA, and where sequence data are available, a direct

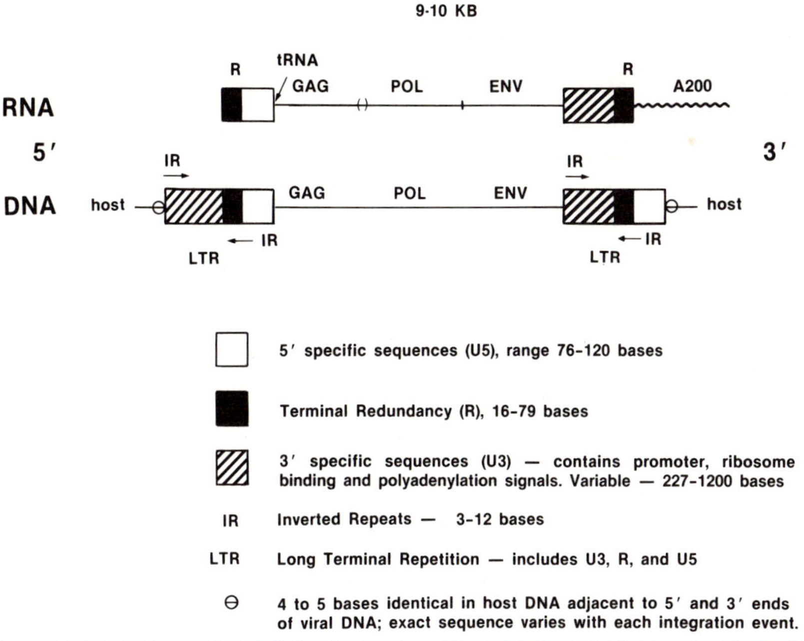

Figure 5 Structure of viral RNA and integrated proviral DNA. (From H. M. Temin. *Chest 27*, 1-3, 1981.)

repetition of four to five bases of host DNA is found adjacent to each end of the proviral DNA. Transcription of proviral DNA involves at least two species of mRNA, one for *gag-pol*, which is of genome size and must contain stop codons after *pol*, and the second for *env*, which contains a leader sequence from the 5'-region of the genome spliced to the *env* transcript.

Primary Structural Analyses of Type C Virus Proteins

As noted above, the mammalian p30s have shown cross-reactive (interspecies) determinants, as established by both immunodiffusion and radioimmunoassays, and they have been found to be compositionally similar. Sequence analysis performed by Dr. S. Oroszlan (NCI-Frederick Cancer Research Facility) provided support for the immunologic findings and showed that the NH_2-terminal regions of p30s of mammalian type C viruses are highly conserved. Four major subgroups have been identified based on the alignments of amino terminal sequences of p30s. This classification is shown in Figure 6. All p30s initiate with a common tripeptide, namely, prolylleucylarginine, and contain a highly homologous (conserved) region beginning with glutamine at position II in the sub-

	1 5 10 15 20 25 30
Subgroup I[a]	
Mouse	
R-MuLV	P L R L G G N G Q L Q Y W P F S S S D L Y N W K X D N P A F
Cat	
FeLV	P L R E G P N N R P Q Y W P F S A S D L Y N W K S H N P P F
Subgroup II[b]	
Baboon	
BaEV	P L R T * V N R T I Q Y W P F S A S D L Y N W K T H N P X F
Cat	
RD-114	P L R T * V N R T V Q Y W P F S A S D L Y N W K T H N P X F
Subgroup III[c]	
Gibbon ape	
GaLV	P L R A I G N G P L Q Y W P F S X A D L Y
	inserts: (P P A E P) at the arrow after G; (L V) at the arrow after the second G
Subgroup IV[d]	
Rhesus	
MMC-1	P L R E I G T G R L M Y W P F S T S D L Y N W K
	inserts: (S L D D) at the arrow after G; (L H) at the arrow after the second G

Figure 6 Classification of mammalian type C viruses based solely on alignment of NH_2-terminal sequences of p30s. The classification shown here is in general agreement with that derived from other comparative methods. Here emphasis is given to gaps or inserts as major events in the evolution of homologous proteins.

[a]Mouse p30s from at least 7 isolates have identical sequences through 30 residues with the exception of position 4 where L, S, or A have been found. Note that FeLV aligns without gaps/inserts even though there are sequence differences. Viruses of rat and hamster origin also align in this fashion.

[b]Baboon and RD-114 viruses require a gap (logically assigned at position 5) to maintain alignment with Subgroup I. Note that only a single difference exists (position 10, including gap) between these terminal sequences.

[c]The woolly monkey virus p30 is identical to the GaLV sequence. Note that 7 residues are inserted to maintain alignment with Subgroup I.

[d]The sequence of the p30 of a virus isolated from *M. arctoides* (MAC-1) is virtually identical to this sequence; also, a new isolate from colobus monkeys shows a high degree of homology and aligns without gaps/inserts. Six residues must be inserted to maintain alignment with Subgroup I.

Note: The one letter code used for amino acids is: A, alanine; D, aspartic acid; E, glutamic acid; F, phenylalanine; G, glycine; H, histidine; I, isoleucine; K, lysine; L, leucine; M, methionine; N, asparagine; P, proline; Q, glutamine; R, arginine; S, serine; T, threonine; V, valine; W, tryptophan; X, unidentified; Y, tryosine; *, gap. (These data were derived from studies of S. Oroszlan, NCI-Frederick Cancer Research Facility.)

group I sequences (a few additional substitutions also occur in this region). While in the first subgroup (mouse, rat, and cat leukemia viruses), residues 4 to 10 are variable and essentially species specific, the p30s in subgroup II (baboon endogenous virus [BaEV] and cat endogenous virus [RD-114]) differ only in position 9 and both have a deletion in position 5. Gibbon ape leukemia virus (GaLV) and simian sarcoma-associated virus (SSAV) in subgroup III have identical sequences throughout the known first 28 residues, and both have two inserts involving a total of seven residues in the variable region of the mammalian p30s. In subgroup IV, the p30s of the recently isolated endogenous rhesus monkey (*Macaca mulatta*) virus, MMC-1, and that of *Macaca arctoides*, MAC-1, are also identical (one possible difference remains to be verified), but have a six-residue insert which is divided in the alignment into two inserts, one involving four and the other two residues. These sequences have the first residue glutamine in the conserved region substituted with methionine. The NH_2-terminal amino acid sequences of mammalian p30s and the structural homolog p27 of endogenous avian type C viruses show no obvious sequence homology. Reliable serologic analyses, including highly sensitive radioimmunoassays, have not revealed antigenic cross-reactivity between *gag* gene products of endogenous viruses of the avian leukosis-sarcoma group and of mammalian type C viruses. Due to inherent limitations of immunoassays, the lack of immunologic cross-reactivity does not mean that avian and mouse type C viruses are not genetically related. Indeed, we have predicted that homology in primary structure should exist between the *gag* gene products of avian and mammalian oncornaviruses. This is based on the assumption that contemporary viral genes are descendants of a common ancestral gene that has undergone divergence in the various viral species. Since mouse p10 is highly conserved (based on the sequence analysis of several mouse p10s) and avian p12 is considered to be a functional homolog, a comparison of the sequences of these proteins should provide a sensitive measure of interviral relationship. A recent statistical analysis of these sequences suggests a definite evolutionary relationship between these avian and mouse type C virus *gag* gene-encoded proteins.

In other sequence homology studies, Oroszlan and colleagues found that the NH_2-terminal amino acid sequences of mouse leukemia virus pp12 showed definite homology with the NH_2-terminal conserved region of goose and chicken H5 histones, the phosphorylated nuclear proteins of nucleated erythrocytes. The implication of these findings is that the conserved structural feature may indicate a common function. Both the p12 and H5 proteins bind specifically to nucleic acids through interactions modulated by phosphorylation. The sequence homology between type C virus phosphoproteins and H5 histone may represent the first example of usurpation of a normal cellular transcriptional unit for use in a retrovirus.

This section has provided details of only a few primary sequence comparisons. These analyses have been extended to other virion proteins, and combined with DNA sequencing data, they will permit complete descriptions of virion proteins. Similar information is also becoming available for transformation-related proteins encoded by the *src* and analogous genes.

CONCEPTS AND SPECULATIONS

The wide range of pathologic, virologic, and molecular biologic studies of the type C viruses have generated speculations and concepts that have bearing on the general fields of biology and medical science, particularly in regard to mechanisms of normal and neoplastic cell growth. A number of these are now well accepted and are included in our current concepts of molecular biology; others are still merely interesting speculations and hypotheses.

Origin of Type C Viruses

Sir MacFarland Burnett postulated that the ultimate origin of viruses might be from prokaryotic microorganisms or cellular genetic information of eukaryotic organisms. The accumulated information regarding type C viruses suggests that they may have evolved from cellular genes. Nucleotide sequences related to the type C viral genes, *gag*, *pol*, and *env* (virogenes of Huebner and Todaro), now appear to be a family of cellular DNA proviruses or virogenes in both avian and mammalian species. These nucleotide sequences have been detected in multiple copies in the genomes of cells of chickens and other avian species and of mice, rats, hamsters, cats, pigs, baboons, and other primates. The method of regulation of these genes is unknown, although Temin has proposed that a cis-acting control element adjacent to the DNA provirus may regulate their expression. It has been observed that normal or cancer cells sometimes spontaneously release these type C viruses and also that they can be induced to release virus by a number of methods, including prolonged culture in vitro, treatment with halogenated pyrimidines or inhibitors of protein synthesis, graft-versus-host reactions, or infections with foreign type C viruses. It appears that nontransforming type C viruses exist in nature in two forms: as DNA proviruses constituting a family of cellular genes and as extracellular type C viruses with an RNA genome. In addition, evidence is accumulating that retroviral DNA sequences constitute a substantial portion of the mammalian genome. In the most thoroughly studied organism, the mouse, at least four different classes of endogenous retroviral-related sequences are present, accounting for up to 0.35% of the mouse genome. In certain mouse species this includes six copies of Gross-type murine leukemia virus, six of wooly monkey and gibbon ape viral-related genomes, up to 25 of *Mus cervicolor*, M432, viral genomes, and at least 25 mammary tumor viral-related type B viral genomes. Also present in several strains of mice are 400 to 1800 copies of intracisternal A particle sequences related to the M432 genome. In primates, six different endogenous retroviruses have been detected and characterized. These include owl monkey type C and squirrel monkey type D viruses (new world monkey retroviruses) and baboon, macaque, and colobus type C and the Langur monkey type D viruses (old world monkey retroviruses). In each case, the retrovirus is integrated in the host genome in multiple copies. In the baboon, multiple copies of three different primate retroviruses have been detected thus far. Baboon type C endogenous virus sequences are reiterated about 100 times. The closely related macaque type C viruses, MMC-1 and MAC-1, which have limited sequence homology with baboon virus, are also highly reiterated in baboon cell DNA. Further, the Langur type D virus, which is also unrelated to baboon virus, is homologous to sequences present in baboon cell DNA. Together, these proviruses constitute 0.02 to 0.04% of the baboon genome. Transfection studies with endogenous and exogenous baboon proviruses along with hybridization studies and restriction enzyme mapping of the proviruses, have yielded interesting information about the sequence organization and possible regulation of the baboon type C retroviral genes. In general, despite the approximately 100 integrated baboon virus-like genomes in baboon cell DNA, only a low level of infective virus is released from some placental or embryonic cells, and transfection with baboon DNA, even when positive, is inefficient. This is not unexpected, because the multiple integrated viral genomes contain considerable sequence heterogeneity. By comparison, DNA isolated from human cells productively infected with baboon virus and releasing high titers of virus yet containing only exogenous proviral copies is highly infectious by transfection. In addition, hybrids formed between baboon cell DNA and labeled baboon virus complementary DNA were found to have a lower thermal stability and a broader melting profile than hybrids formed between the human producer cell DNA and the same probe.

Restriction enzyme mapping, performed by Dr. M. Cohen (NCI-Frederick Cancer Research Facility) has been made of the unintegrated proviral genome in acutely infected human cells and integrated proviral genomes in productively infected human cells and in baboon tissues. Of considerable interest is the observation that in contrast to the integrated endogenous proviral DNAs in baboon tissues, the unintegrated provirus and integrated provirus in human cells have a 150 nucleotide base pair deletion. From these early observations it is tempting to speculate that the specific deletion coincides with provirus activation and renders the DNA infectious in transfection studies.

In relation to the origin of these endogenous nontransforming mammalian type C viruses, it would appear that on the basis of immunologic and nucleic sequence analysis most of them can be traced back to three main lineages of ancestral viruses: one of rodent origin giving rise to the mouse, rat, hamster, cat, pig, woolly monkey, and gibbon ape viruses; the second, of primate origin, giving rise to the baboon and endogenous cat RD-114 viruses. The third group contains the endogenous viruses of macaques (MMC-1 and MAC-1) and the related isolate from the colobus monkey. In addition to these groups of nontransforming replicating type C viruses, the origin of the replication-defective transforming (sarcoma) viruses must be considered. These viruses appear to have arisen as a result of genetic recombination between the nontransforming viruses and other host cell genetic sequences (the cellular *sarc* sequences). Mammalian type C sarcoma virus isolates have been restricted to four species; these include two rodents (mouse and rat), one carnivore (cat), and one primate (woolly monkey). Of interest, a stable rat sarcoma virus has recently been isolated in the laboratory as described previously. Hitherto, sarcoma viruses had been derived from sarcomas occurring in nature or after transplantation of sarcoma cells or type C nontransforming viruses in laboratory animals. It would now appear that in the avian system, chicken cells contain not only *sarc* sequences associated with the production of type C sarcoma viruses, but also *carc* sequences leading to the production of type C viruses with *gag*, *env*, *pol*, and *carc* genes that can cause experimental carcinoma in host animals. In addition, viruses inducing specific tumors of the myeloid and erythroid systems have been isolated and found to contain unique transformation-related genes.

The findings noted above of the sequence homology between viral and cellular proteins is consistent with the emergence of infectious entities by a process of differentiation from preexisting cellular genes. Vestiges of these primordial origins might be more readily detected in retroviruses because of their inheritance in the cellular genome. Shifts in the physical location of specific DNA segments similar to those that occur during maturation of the immune system might have contributed to the origin of viral structural proteins. In this manner, several cellular genes or coding domains of diverse function could have contributed to the origin of individual viral polypeptides. Further analyses of the complete sequences of retroviral polypeptides now becoming available will be necessary to evaluate these speculations.

As also noted earlier, transforming viruses appear to have arisen by a recombinational event between nontransforming helper virus and host DNA and that it is this excised segment, *sarc* or equivalent, of host DNA which, when exogenously introduced into an appropriate cell, results in transformation. It has indeed been found that the normal sequence in mouse DNA which is related to the *src* region of mouse sarcoma virus can be isolated and that it carries transforming potential. This leads to the hope that a similar array of sequences in human DNA might be identified which, when removed from normal control processes, contributes to neoplasia. One approach to this possibility will take advantage of the emerging generality that genes related to the transforming regions of transforming viruses are highly conserved and can be detected across wide taxonomic distances. Thus, transforming viral probes might be used to identify and isolate regions of human DNA involved in cell transformation.

The Viral Oncogene Hypothesis

This hypothesis, now of historical interest in view of the concepts discussed above, was first proposed by Huebner and Todaro in 1969. The hypothesis stated that all vertebrates contain the genetic information for producing a type C RNA virus in both somatic and germ cells. This information has been part of the genetic makeup of vertebrates since early in the evolutionary process. The virogenes (genes for production of type C viruses) and the oncogenes (that portion of the genome responsible for transforming a normal cell into a tumor cell, now *sarc* gene) are maintained in an unexpressed form by repressors in normal cells. Various agents, including physical and chemical carcinogens, may transform cells by "switching on" the endogenous oncogenic information. This hypothesis was based on evidence indicating a general lack of overt virus expression in species (mainly chickens and mice) where the viral genome was known to be present, either based on detection of viral antigens in embryos or ability to induce virus from cultured cells by a number of physical or chemical treatments. Considerable support for the hypothesis came from molecular hybridization experiments indicating the presence of the genome in cell DNA. Current evidence indicates that a rigid conservation of the virogenes does not occur and, frequently, closely related species may not contain expected sequences. There is clear evidence for both evolutionary divergence and interspecies transfer as interacting mechanisms impacting the current distribution of retroviruses and related genes. In contrast, the oncogene portion of the hypothesis remains viable as the evidence indicates a high degree of evolutionary conservation of those genes associated with viruses causing acute malignant conversion.

The Protovirus Hypothesis

This hypothesis, also of historical interest, was proposed by Temin after the discovery of the RNA-dependent DNA polymerase. Because transformation of chicken cells by Rous sarcoma virus induced a genetically stable cellular change and required early DNA synthesis and because Rous sarcoma virus production was sensitive to actinomycin D, Temin proposed that Rous sarcoma virus replicated through a DNA intermediate, the provirus. Because the formation of the DNA provirus was not sensitive to inhibitors of protein synthesis, Temin proposed that the virion must contain an RNA-directed DNA polymerase which was responsible for transferring information from viral RNA to the DNA provirus. Following these observations, Temin proposed the protovirus hypothesis for the etiology of cancer. This hypothesis suggests that, for transformation of cells by Rous sarcoma virus, there must be formation in the cell, by means of the virion RNA-dependent DNA polymerase, of genes for viral replication and transformation. The fact that the expression of these two sets of genes is separate is shown by the existence of virus-producing nontransformed cells and of transformed cells that do not produce virus.

In the protovirus hypothesis, in contrast to the oncogene hypothesis, the genetic information required for cell transformation does not exist in the germ cells. Here, a normal process of DNA→RNA→DNA information transfer, operative in development, is deranged to give rise to the formation of the genes for neoplastic transformation. Current evidence indicates that genes highly related (perhaps identical) to those found in transforming viruses do exist in normal DNA; however, some maturation process may still be operative before these genes become active.

Transfer of Viral Genes between Species

Because RD-114 virus was isolated from human rhabdomyosarcoma cells after transplantation in fetal cats, extensive studies were made to characterize and identify the virus and to determine its role, if any, in feline and human cancer. During these studies, a number of interesting observations were made. RD-114 virus was identified as a xenotropic, endogenous virus of domestic cats and three closely related feline species (*Felis sylvestris*, *F. margarits*, *F. chaus*). Further, a curious relation between RD-114 and the endogenous baboon type C viruses was discovered. When the first baboon virus (M-28) was isolated, it appeared to be similar to RD-114 virus in biologic and immunologic parameters, and it was originally considered to be either a RD-114 virus contaminant or a baboon virus related to RD-114; the latter possibility proved to be correct. RD-114 and the baboon viruses have a similar host range, induce syncytia, and show reciprocal interference and a degree of cross-neutralization in tissue culture. The p30s of both viruses show identity reactions in gel diffusion using sera nonreactive with other mammalian type C viruses and their NH_2-terminal amino acid sequences are identical for the first 14 residues. Further similarities are the phosphorylation of the pp15 in both viruses as compared to phosphorylation of the pp12 in other mammalian type C viruses; also, the DNA primer is tRNA gly compared to tRNA pro in other mammalian viruses. Finally, the baboon virus polymerase is partially inhibited by concentrations of RD-114 antipolymerase sera which do not inhibit enzyme activity of other type C viruses. Molecular hybridization studies, however, show a clear distinction between the genomes of the two viruses which are only approximately 10 to 20% related. The origin of these two viruses was then studied by looking for the presence of the viral genes in tissues of different species. Using stringent hybridization conditions, the cellular DNAs for old world monkeys and domestic cat tissues were found to contain nucleic acid sequences partially related to endogenous baboon viral sequences and, reciprocally, DNAs from baboon and nine other old world monkeys contained sequences partially related to the RD-114 virus genome. The DNA of other primates (prosimians, new world monkeys, apes, and humans), other wild Felidae species, or other mammals (mouse, rat, mink, cow, sheep, pig, and dog) did not appear to contain such sequences or gave minimal reactions, i.e., 8% with chimpanzee DNA. Using less stringent conditions, significant reactions (about 20% hybridization) were obtained with DNAs from the higher apes, whereas prosimians, new world monkeys, and nonprimates (except the domestic cat) remained negative. These observations suggest that the endogenous baboon virus is a contemporary representative of type C gene sequences present in old world monkeys and apes for most of their evolutionary history and that the presence of a related genome in the domestic cat, but not all Felidae, cannot be explained on evolutionary grounds. Todaro's group theorized that the progenitor virogenes were of primate origin and spread to the cat because of their presence in the DNAs of old world monkeys and apes, indicating their long-term residence (at least 40 million years) in the primate line. The absence of virogenes in the DNAs of most Felidae species, except the domestic cat and a few close relatives, suggested a more recent spread to this line.

These observations were the first to suggest that type C viral genes could be transferred, apparently under natural conditions, between species only remotely related phylogenetically. Thereafter, two additional examples of gene transmission were theorized. First, that the cat leukemia virus (FeLV) was transmitted from an ancestor of the rat to ancestors of the domestic cat and their close relatives. The relationships observed between FeLV and the endogenous viruses of rodents are similar to those between the endogenous RD-114 feline viruses and the endogenous baboon primate type C viruses. FeLV-related sequences are found not only in the cellular DNA of domestic cats, but also in the DNA of the same other closely related Felidae, as in the case of baboon virus. Other more distantly related *Felis* species lack FeLV-related virogenes, while the cellular DNAs of rodents, rats in particular, contain related virogene sequences. This was interpreted to suggest that FeLV-related genes were introduced into the *Felis* lineage following transspecies transfer by type C viruses of rodent origin. It was of interest that the four species of cats containing RD-114-related genes also contain FeLV-related genes, while other *Felis* species lack both sets of sequences. The third example of gene transmission between species is from an ancestor of the mouse to an ancestor of the domestic pig. Pig type C viruses are genetically transmitted, and the viral-related sequences are present in multiple copies in pig cellular DNA. Close relatives of the domestic pig, the African bush pig and European boar, have closely related pig virogene sequences in their DNAs. It was suggested that the virus was acquired by an ancestor of the pig from a small rodent related to the mouse. The nucleic acid homology between the endogenous pig type C viral RNA and murine cellular DNA suggests that the virus had a murine origin. From the extent of hybridization of the pig type C viral DNA probes to rodent cellular DNA, it was concluded that the virogenes in the pig were transmitted from members of the Muriadae family after the mouse had separated from the rat and before different species of mouse had diverged from each other. These three examples demonstrate that viral genes from one group of animals might integrate into the DNA of animals of another species and become incorporated into the germ line. It has been hypothesized that, if viral gene sequences can be acquired in this way, it is possible that type C viruses have served to introduce other gene sequences from one species to another and may provide a mechanism by which species can stably acquire new genetic information.

Use of Type C Viral Genes in Evolutionary Studies

Implicit in the concept of interspecies transfer of viral genes is the idea that endogenous type C viral-related sequences can be used as evolutionary markers. Todaro and Benveniste proposed that, if the baboon type C viruses were in fact endogenous primate viruses, then it would be logical for other old world monkeys who are close relatives of the baboon to have related sequences. The prediction was that a decreased extent of hybridization and a lower thermal stability of the duplexes formed when the heterologous cellular DNA was hybridized to the baboon viral complementary DNA probes. Studies of evolutionary relationships of type C viral sequences were favorable to conduct in primates since so much is known about their evolutionary relationships. The old world monkeys, which include the baboon species, diverged from the greater apes and humans 30 to 40 million years ago. The new world monkeys diverged from the common primate stem leading to old world monkeys and apes approximately 50 million years ago, while prosimians evolved from the primitive mammalian lineage 60 to 80 million years ago. Hybridization studies using baboon type C virus gave positive results in all old world monkeys, and the extent of hybridization was directly related to their taxonomic closeness to baboons. Sequences were also found in apes, especially chimpanzees and gorillas. No sequences were found in new world monkeys or prosimians. These results apparently indicate that the type C viral genes evolved just as the primate species evolved, with virogene sequences from more closely related genera and families showing more sequence homology than those from more distantly related groups. The presence of endogenous type C viral sequences among anthropoid primates and their conservation for at least 30 to 40 million years indicates that they have evolved in primates over this length of time as stable cellular elements equivalent to other cellular genes. These results also suggested that the viral genes may provide some normal functions advantageous to the species carrying them.

Role of Virus in Normal Cell Function

The presence of genetically transmitted virogenes in various vertebrate species and the evidence that they have been conserved, at least in some cases, through evolution suggests that they may play a role in the host cell. The first role, suggested by Huebner and Todaro, was that the expression of viral information during the early stages of embryogenesis and cellular differentiation in the mouse was a normal and essential part of the developmental process. After being expressed in embryonic tissue, the p30 antigen in some mouse strains was not expressed after birth. They and other authors noted that the p30 antigen along with complete virus was expressed later in life in mice who developed leukemia and lymphoma. It was suggested that cancer later in life might be related to an inappropriate manifestation of a normal developmental function. Such a normal function would help to explain why they are conserved during evolution. Another type C viral protein, gp70, has been linked to differentiation and development in the mouse. Immunofluorescent studies have demonstrated that the expression of gp70 in certain anatomical sites is found to be higher in lymphoid and epithelial cells than in other cells. The major site of gp70 expression was reported to be the male genital tract, and a protein immunologically and biochemically related to gp70 was found at high levels in the secretions of the epididymus and ductus deferens. Further, the results of other studies have established that gp70 is a constituent of the surface of normal thymocytes and that it shares their immunologic and biochemical properties with the thymocyte differentiation marker GIX. One serious objection to this area of speculation is a recent discovery that chickens may be freed of all endogenous proviruses by selective breeding without detrimental effects on viability.

Another possibility is that type C viruses may have played an important evolutionary role as transmitters of genetic information not only between cells of an animal and individuals of a species, but also between species. As noted above, the viruses might transmit themselves between the germ line DNAs of different species; this has been demonstrated experimentally.

CONCLUDING COMMENTS

The initial demonstrations that retroviruses could exist as inherited elements within the cell genome led to speculations of their role in the normal development of the host species. Interspecies transmission was also seen as a means of enriching the genetic variability of new hosts. Such ideas added to the substantial interest in these viruses generated by their demonstrated oncogenic potential. Despite the novel appeal of a role in normal development, no substantive evidence for this has appeared, and indeed, the extent of heterogeneity between and within a species makes such ideas rather doubtful. It seems more logical to approach this family of viruses from a parasite-host standpoint, utilizing the new information gained from studies of the structural organization of the genome and new concepts allowing for independent existence of DNA segments designated "selfish DNA." Thus, for the retroviruses whose integrated state has been studied by direct sequence analysis, the analogy to bacterial transposons is quite clear (Fig. 5). The presence in certain species of inherited elements reiterated from 100 to 1000 times can be explained by replication-dependent transposition, the only selection process being for lack of insertion into vital cell genes causing lethal mutations. As the new genome evolved with time, one would expect to find conservation of the control signals found in the insertion sequence-like elements (LTR) flanking the unique viral structural genes. Variable loss and potential gain of new host sequences within the transposable unit could also occur, accounting for the origin of acute transforming viruses and the occurrence of defective integrated genomes. As described for the baboon endogenous virus, an outbred population would show considerable heterogeneity in number, organization, and function (in terms of making an infectious virus) of its integrated proviruses. If transposition can occur without an infectious cycle, then the only functions which need be retained are those assuring retention and transposition. The final degree of reiteration which can be achieved without detriment to the host may be substantial, as indicated by the A particle genomes in mouse cells which account for at least 0.3% of the cell genome. While we have some confidence that the transposon model will account for the existence of large numbers of proviral DNA copies in normal cells, the acquisition of genes from the host cell leading to transforming potential requires further consideration.

Current evidence indicates that the normal host-transforming genes (*sarc* and equivalents) are found in host DNA apart from retroviral elements and, in this form, are negative in DNA transformation assays. We assume that they are specifically functional during some phase of differentiation, although this remains to be shown. Once spliced appropriately to retroviral elements and containing at least one LTR, the same sequences are active. Since the LTR contains splice signals (mRNAs for *gag* and *env* genes contain these sequences), the generation of transforming viruses would require (1) a splice acceptor sequence adjacent to cellular *sarc*, and (2) insertion of a provirus (or LTR) in a configuration which permits formation of a new transposon including *sarc*. While such events may be rare, it seems possible to design experiments which would use the retroviral LTRs in attempts to rescue cellular genes which can be assayed in selective systems. This approach may be the one of choice in searching for human transforming genes analogous to those described above for the acute leukemia and sarcoma viruses. Finally, retroviruses provide a variety of interesting and unique features for studies of viral origin, interrelationships, evolution, and pathogenesis. These have relevance to a broad set of problems rather than being strictly limited to interest by virologists.

SUGGESTED READINGS

Calos, M. P., and J. H. Miller. Transposable elements. *Cell 20*: 579-595, 1980.

Dhar, R., W. L. McClements, L. W. Enquist, and G. F. Vande Woude. Nucleotide sequences of integrated Moloney sarcoma provirus long terminal repeats and their host and viral junctions. *Proc. Natl. Acad. Sci. USA 77*:3937-3941, 1980.

Doolittle, W. F., and C. Sapienza. Selfish genes, the phenotype paradigm and genome evolution. *Nature 284*:601-603, 1980.

Gilden, R. V., S. Oroszlan, H. A. Young, N. R. Rice, M. A. Gonda, M. Cohen, and A. R. Rein. Genetics of retrovirus-host interactions. In *Frontiers in Immunogenetics* (W. H. Hildemann, Ed.). Elsevier-North Holland, Inc., New York, 1981, pp. 191-223.

McAllister, R. M., M. B. Gardner, M. O. Nicolson, and R. V. Gilden. RD-114 virus: Characterization and identification. In *Progress in Experimental Tumor Research*, Vol. 22 (F. Homburger, Ed.). S. Karger, Basel, 1978, pp. 196-215.

Orgel, L. E., and F. H. C. Crick. Selfish DNA: The ultimate parasite. *Nature 284*:604-607, 1980.

Shimotohno, K., S. Mizutani, and H. M. Temin. Sequence of retrovirus provirus resembles that of bacterial transposable elements. *Nature 285*:550-554, 1980.

Stephenson, J. R. (Ed.). *Molecular Biology of RNA Tumor Viruses*. Academic, New York, 1980.

chapter 8

DNA Tumor Viruses

DAVID T. KINGSBURY

College of Medicine,
University of California, Irvine
Irvine, California

The hypothesized infectious nature of the leukemias of fowl, originally proposed by Ellerman and Bang in 1908, stood for many years as an isolated and unconfirmed observation. The subsequent recognition by Shope in 1933 that the papillomas (warts) of wild cottontail rabbits could be transmitted by cell-free filtered extracts was the second major contribution to viral oncology. We now recognize the Shope papilloma virus as the first example of the DNA tumor viruses.

As the study of tumors of animals expanded, many additional examples of virus-induced tumors were identified. One striking fact about the recognized tumor viruses is that all the oncogenic RNA viruses are members of one group, the retrovirus family, while the DNA tumor viruses are found in every family except the parvoviruses. Table 1 lists the different DNA virus families and some of the recognized oncogenic members of each.

A major development in the study of DNA tumor viruses was the discovery reported in 1953 by Ludwik Gross that some stocks of murine leukemia virus (MLV) contained an activity that induced adenocarcinomas, predominantly in the parotid gland. Gross was able to separate this activity from his MLV stocks by heat inactivation of the MLV. Subsequent studies by Steward and Eddy and others clearly demonstrated that while the parotid tumors were very common the virus also induced a variety of other murine tumors. This ability to transform many cell types led to the suggestion that the virus be referred to as polyomavirus, a name that has been used ever since.

Table 1 Oncogenic Members of the DNA Virus Families

Viral family	Oncogenic members[a]
Parvoviruses	None
Papovaviruses	Polyoma, SV40, SV40-like human papovaviruses, papillomaviruses
Adenoviruses	Many of the recognized serologic types
Herpesviruses	Marek's disease virus, Lucke frog virus, herpes simplex virus, Epstein-Barr virus, cytomegalovirus, primate herpesviruses, guinea pig herpesvirus
Poxviruses	Fibroma virus, yaba monkey virus

[a]Measured by tumor induction in vivo or cell transformation in vitro.

The subsequent work on polyomavirus was pioneering in the field of DNA tumor virus biology and biochemistry. The establishment of an efficient cell culture system for propagation of the virus was followed by a system of in vitro cell culture transformation. These two systems opened the way to quantitative experimental study of the role of the virus in the transformation process. These early studies likewise demonstrated that structurally the virus resembled closely the Shope virus and that the two shared several biologic properties.

An additional contribution to the importance of the papovavirus group in experimental tumor virology was the discovery by Sweet and Hilleman in 1960 of a contaminating simian papovavirus in many lots of poliovirus vaccine. This virus, designated simian virus 40 (SV40), has been intensively studied ever since its discovery due to the finding that it induced tumors when injected into newborn hamsters. The rapid development of excellent cell culture systems for propagation of the virus and in vitro transformation has helped make it the most widely studied of the DNA tumor viruses.

ROLE OF CULTURED CELLS IN DNA TUMOR VIRUS RESEARCH

The development of virology as a discipline has been closely associated with the development of improved methods of in vitro cultivation of cultured animal cells. In no other area of virology has cell culture technology been more important than in studies of the oncogenic viruses. This in vitro system has provided the necessary quantitative basis for the examination of cell-virus interactions.

Oncogenic viruses are able to induce changes in certain cell cultures which lead to a stable alteration of their growth as well as several other biologic and biochemical characteristics. This cell alteration, called transformation, may result from other factors in addition to oncogenic viruses, but the essential features of this process appear to mimic in most ways the process of cell alteration seen in most cancers in vivo.

Many of the properties of transformed cells generally absent from normal untransformed cells are listed in Table 2. It is obvious upon examination of this list that many of these factors are interrelated and that the change in organization of the membrane or cytoskeletal structure will likely lead to changes in cultural behavior, such as anchorage dependence. In the past few years several laboratories have demonstrated that the transformed phenotype in cell cultures can be achieved independently of the cells becoming malignant. That these may be progressive stages on a single pathway seems likely, but both the mechanisms involved in this multistep progression and the various stimuli required to drive this progression remain a mystery.

Table 2 Selected Properties of Transformed Cells Not Present in Resting Untransformed Cells

I. Growth in culture
   - A. High saturation cell density
   - B. Decreased serum requirement
   - C. Less ordered cell growth
   - D. Anchorage independence—growth in agar or Methocel
   - E. Increased cloning efficiency
   - F. Absence of aging

II. Cell surface
   - A. Increased agglutinability by plant lectins
   - B. Decreased large external transformation-sensitive (LETS) protein (fibronectin)
   - C. Presence of new antigens, frequently of fetal origin
   - D. Increased rate of transport activity (nutrients, ions)
   - E. Loss of some hormone and toxin receptors

III. Biochemical alterations
   - A. Disaggregated microfilaments and myosin
   - B. Decreased intracellular cyclic nucleotide levels
   - C. Abnormal collagen synthesis
   - D. Increased aerobic glycolysis

Although most DNA tumor viruses will infect a wide variety of cells, generally only a limited number of cell types are readily transformed. In some cases (i.e., herpes simplex virus) the cytovirulence of the virus demands that inactivated virus be used for transformation regardless of the cell type. The most clearly defined cell-virus relationships exist for the papovaviruses, polyoma and SV40.

Permissive, Semipermissive, and Nonpermissive Host Cells

While both polyoma virus and SV 40 interact with many different cell types, only a few of these cells will support a full productive infection. In this case a productive infection is one in which viral DNA replication and capsid assembly occur with the eventual production of infectious virions. Examples of permissive cell types for both polyoma and SV40 are shown in Table 3. An example of each is primary mouse embryo cultures permissive for polyoma virus and cultures of African green monkey kidney permissive for SV40. In contrast to the completely permissive cell, the nonpermissive cell survives an infection by these viruses and no progeny virus particles are produced or released. In some of these cells a limited number of viral genes are expressed but always without the production of virus particles.

These two cell types constitute the ends of a spectrum between which lie cells which are termed semipermissive. When populations of these cells are challenged some members are infected in a permissive

Table 3 Examples of Cell Types Permissive, Nonpermissive, and Semipermissive for SV40 and Polyomavirus

Virus	Cell class	Type of culture
SV40	Permissive	Primary cultures of African green monkey kidney; continuous African green monkey cell lines (Vero, BSC-1)
	Semipermissive	Human fibroblasts; hamster kidney
	Nonpermissive	Mouse 3T3 cells; mouse fibroblasts
Polyoma	Permissive	Mouse embryo cells; mouse 3T3 cells
	Semipermissive	Hamster embryo cells; BHK cells (baby hamster kidney)
	Nonpermissive	Human fibroblasts; African green monkey cells

manner and die, accompanied by the production of viral particles, whereas others appear nonpermissive for virus replication and survive the infection, frequently with associated transformation. The reasons for this semipermissiveness in certain cell populations remain unknown; however, experiments on mouse-hamster cell hybrids have suggested that the expression of certain specific genes may control the relationship between an infecting virus and the host cell.

PAPOVAVIRUSES

General Properties

The papovavirus group is one of the most well understood in animal virology. These are small naked capsid viruses, and almost all members of the group are identical in size and structure. They are icosahedral in symmetry and the protein coat contains 72 capsomeres. Both polyoma and SV40 are 45 nm in diameter, while the papillomaviruses are slightly larger at 55 nm in diameter.

The 72 capsomeres of the viruses are made up of 420 structural units produced from three structural polypeptides. These major capsid proteins, designated VP1, VP2, and VP3, account for 70 to 80% of the total protein of the virions. VP1 is the major structural protein of the virus and it alone accounts for 75% of the virion protein. This polypeptide ranges in molecular weight from 47,000 for polyoma to 45,000 for SV40. The structural proteins VP2 and VP3 have molecular weights of 35,000 and 23,000 in polyoma and 42,000 and 30,000 for SV40. The remainder of the virion-associated protein consists of a series of histones associated with the viral DNA.

One unusual property of this virus group is the presence of the viral DNA in the form of chromatin rather than naked DNA. This feature is a consequence of the association of specific histones with the viral DNA during replication. These histones, H3, H2A, H2B, and H4, are known to be of host origin; however, they are extensively modified by acetylation during virus infection.

The DNA of the papovaviruses consists of a covalently closed circular molecule with a molecular weight of 3.4×10^6 for polyoma and SV40 and 5×10^6 for the papillomaviruses. Within the virion the DNA is in a highly structured form and complexed with histone proteins. After removal of the protein the DNA is isolated as a superhelical structure due to its closed circular nature. This supercoiled form of the DNA is referred to as form I and it is converted to the open circular (form II) configuration by the introduction of a single nick which allows the two strands to move relative to each other and relieve the superhelical tension.

In the past few years, following the introduction of rapid DNA sequencing techniques, the complete nucleotide sequence has been determined for both polyoma and SV40, as well as one of the human papovaviruses, BK virus. SV40 consists of 5226 nucleotide pairs, and polyoma virus, 5292.

Lytic Cycle of Polyoma and SV40

The infection of mouse cells by polyomavirus or green monkey cells by SV40 initiates a complex series of events eventually resulting in the production of thousands of progeny virus particles and the death of the infected cell. The sequence of these events is a carefully controlled and regulated process. Virus infection results in both the induction of new viral gene products and the stimulation of cellular enzyme and DNA synthesis.

The first obvious feature of the lytic infection process is the strict temporal sequencing of the infection events. Following adsorption to the cell surface the virus penetrates the cell membrane and finds its way to the nucleus where replication begins. The mechanism of this penetration-uncoating-migration process remains one of the least understood features of papovavirus replication.

Expression of the viral genome begins in the nucleus, and after 10 to 12 h of infection the first of the virus-encoded proteins appears. Simultaneous with the synthesis of viral gene products is the stimulation of host cell metabolism. This stimulation is generally associated with inducing the movement of the cell into a portion of the cell cycle (G1) leading to DNA synthesis (S). The effects on cellular enzymes include an increase in the level of DNA polymerase as well as the enzymes involved in the biosynthesis of pyrimidine deoxyribonucleosides. In addition, the rate of ribosomal RNA and overall protein synthesis also increases as much as 20 to 30%. After 12 to 15 h of infection, both viral and cellular DNA synthesis are initiated. Following viral DNA synthesis a new class of virus messenger RNA appears, followed by the synthesis of virus capsid proteins, and by 20 to 25 h of infection the first infectious virus starts to appear. Maturation continues until 60 to 70 h postinfection, at which time the cells die. The detailed sequencing of these events has been the subject of intensive study, especially in light of the fact that many of these processes are similar or identical to events which occur during the early stages of cell transformation.

The expression of the viral genome in permissive cells is clearly separated into two phases, designated early and late. The mechanism of switching from the early to the late transcriptional mode is one of the most widely studied features of papovavirus infection and is clearly regulated at the transcriptional level. It is also clear that the switch from the early to the late transcriptional mode is associated with the onset of viral DNA synthesis.

Early mRNA and Early Gene Products

The early transcriptional patterns in both polyoma and SV40 are well understood despite the fact that only very small amounts of early RNA appear on polyribosomes. The development of strand separation techniques, as well as restriction enzyme mapping of the DNA, has permitted the assignment of exact regions of defined DNA strands to these transcripts.

The demonstration that the primary nuclear transcripts are subsequently processed prior to their appearance on polyribosomes has clarified many of the mysteries of the apparent overlapping sequences of the early transcripts. The likely template for this early transcription is the viral DNA complexed into its chromatin form with the nucleosomal histones. It is very likely that cellular RNA polymerase II is the enzyme responsible for this transcription. Hybridization experiments with bulk nuclear RNA suggest that approximately one-half of the E (early) strand is transcribed and none of the L (late) strand is copied at this time. Following transcription the subsequent processing ("splicing") events generate at least two partially overlapping RNA molecules. The general features of the transcriptional program are outlined in Figure 1, which is a composite genetic and transcriptional map for both polyoma and SV40.

Translation of this early mRNA produces the early gene products, and their structure and function are perhaps the most interesting feature of papovavirus biology. The principal early translation products are nonstructural proteins and are the same proteins found associated with transformed cells and virus-induced tumors. These nonstructural early proteins were originally identified as the T antigens (tumor antigens) detectable in infected cells by immunofluorescence, complement fixation, or immunoprecipitation. More recently, identification of these polypeptides as early gene products has utilized purification of polyribosomal mRNA followed by in vitro translation of that RNA and identification of the translation products.

Both polyoma and SV40 induce the synthesis of a major large polypeptide designated large T antigen, with a molecular weight of 80,000 to 100,000 determined by gel electrophoresis or by guanidine-Sepharose chromatography. In addition to large T, there are at least two other polypeptides synthesized during early infection. Both polyoma and SV40 induce a small T antigen in the molecular weight range of 17,000 to 22,000, and polyoma-infected cells have a middle T antigen of approximately 50,000 to 60,000 molecular weight. The presence of middle T in SV40-infected cells is not certain, but a membrane trans-

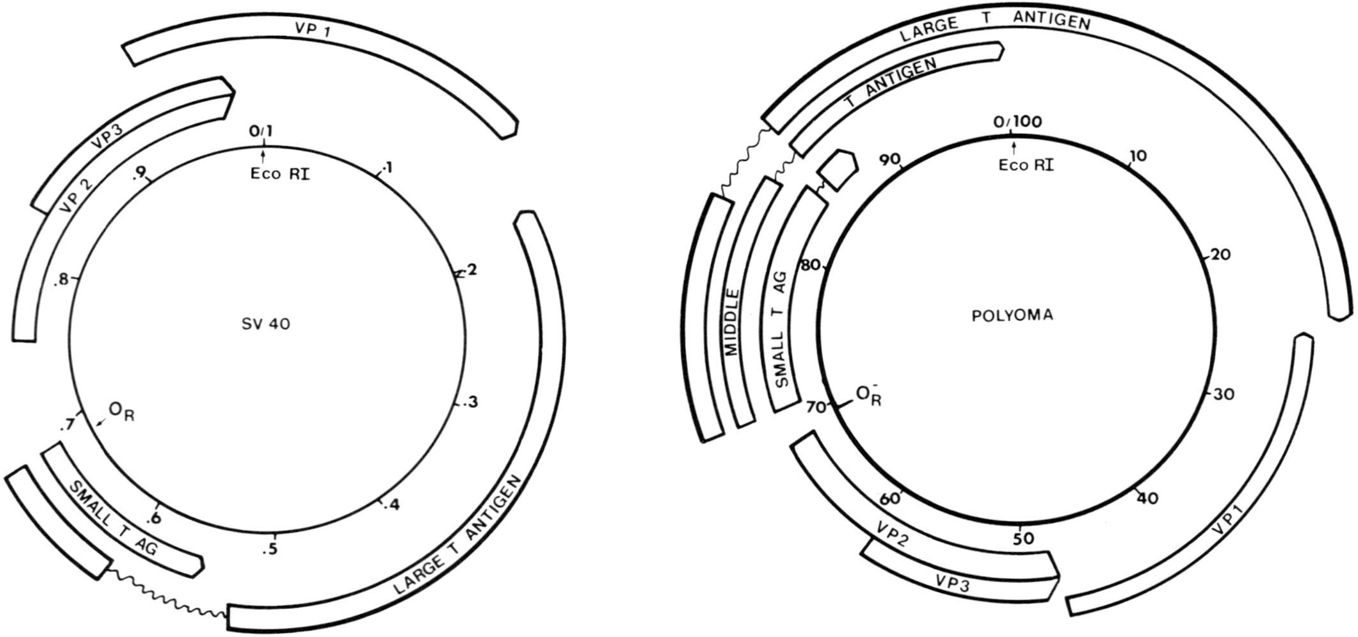

Figure 1 Schematic genetic and transcriptional map of polyoma and SV40. The arrows point in the 5'-3' direction. The boxed areas signify the actual coding regions, and the breaks in the boxes signify the positions of the nontranslated introns. The 5' leader sequences have been omitted as have the 3' tails and poly (A) regions. The numbers on the circular map indicate the fractional map distance from the single EcoRI site present in each molecule.

plantation antigen (TSTA) is present. The sum of the molecular weights of the early polypeptides exceeds the coding capacity of the segment of the virus DNA copied during early synthesis. This fact led to the speculation that large T may not have been a direct viral gene product. It is now quite clear, as diagramed in Figure 1, that all of these antigens are virus coded and result from different "splicing" events during RNA processing.

Large T polypeptide is a multifunctional protein and has several clearly defined activities. (1) It is responsible for the induction of the cellular deoxypyrimidine kinases and cellular DNA synthesis. (2) The large T polypeptide is a regulatory protein for its own synthesis as well as that of other early virus-coded proteins. (3) The presence of active large T polypeptide is required for the initiation of viral DNA synthesis. (4) Large T polypeptide is essential to the establishment and maintenance of transformation (to be discussed below in more detail).

Small T is also a virus-coded early gene product. This polypeptide has identical N-terminal sequences with large T but is not a simple cleavage product of the larger protein. Instead, it is a unique polypeptide which includes, in addition to the short shared nucleotide sequences, a segment of the primary transcript which is "spliced out" during the processing of large T mRNA. Likewise, mutants known to be affected in the small T polypeptide (polyoma *hr-t*) are able to complement polyoma temperature-sensitive mutations in the large T (ts A) polypeptide.

The function of small T in productive infection remains unknown. Mutants of SV40 with deletions in the portion of the viral genome coding for the unique portion of the small T polypeptide (*dℓ* F mutants) replicate at normal or near normal rates although the virus yield may be slightly reduced. Polyoma mutants in the *hr-t* gene (small T polypeptide) grow very poorly in some cell lines but nearly as well as wild-type in others. It appears that the *hr-t* gene product is necessary for permissive growth but that some cell types produce cellular polypeptides which may serve the same functions. There is some suggestion that this activity is associated with histone induction or modification.

In polyoma-infected cells a third early gene product, middle T polypeptide, appears. This antigen is also present in polyoma-transformed cells. SV40 does not appear to induce a polypeptide analogous to middle T during lytic infection. The polyoma middle T polypeptide has an unusual structure relative to the other T polypeptides. Detailed structural studies comparing the transcriptional maps and tryptic digests of the three polyoma antigens have shown that the amino terminal of middle T is derived from sequences shared with large T and that the small T sequences are included together with a limited number of unique sequences. The role of this polypeptide in lytic infection is a complete mystery although it may be required for transformation (discussed below).

DNA Replication

As mentioned above, the switch which defines the change between the early synthetic and the late synthetic period is the onset of viral DNA synthesis. This synthesis begins somewhere between 12 and 15 h following infection of permissive cells. The exact conditions which need to be met in order for viral DNA synthesis to begin are not fully understood, however. The level of stimulation of host cell synthetic enzymes, as well as the level of the large T polypeptide, appear to be among the critical factors. There is good evidence that a certain threshold level of large T is required for the initiation of each round of viral DNA replication, and work with temperature-sensitive mutants has shown that a fully functional large T polypeptide is necessary.

The exact mechanism of the initiation of DNA synthesis by large T is not understood, but in vitro studies with purified SV40 large T polypeptide isolated from cells infected with an adenovirus-SV40 hybrid suggest that it binds specifically in the region of the origin of DNA replication. Following initiation, replication proceeds bidirectionally from the single initiation site. As the daughter strands grow, a larger and larger portion of the parental DNA becomes unwound. However, electron microscopy of replicative intermediates clearly shows that the unreplicated part of the parental molecule retains its superhelical turns, indicating that initiation did not involve the insertion of a permanent nick in the parental molecule. Instead, the unwinding process occurs continually during replication and appears to be mediated by an enzyme, found in normal uninfected cells, which rapidly introduces a nick and reseals it again. During this time both growing points are moving around the circular DNA molecule and eventually meet 180° around the circle from the point of initiation. All the available experimental evidence suggests that there are no unique sequences or any unique structures associated with the termination of replication.

The result of this mode of replication, which accounts for at least 80% of virus DNA synthesis, is the generation of two interlocked circular DNA molecules which must be separated by breaking one of the continuous parental strands. This is probably accomplished by an enzyme like that involved in the unwinding process. After segregation, the two ends are ligated together to reassemble a form I molecule.

Late RNA and Late Gene Product Synthesis

Late RNA is defined as that virus-specific RNA synthesized after the onset of viral DNA synthesis, and all the evidence suggests that most, if not all, of the late transcripts are produced from the recently replicated DNA. Late mRNA has as its principal result the production of the capsid proteins necessary for the encapsidation of the virus DNA molecules. Unlike the levels of the early RNA species which are present in very small quantities, the late mRNA is found in much higher amounts making up as much as 0.13% of the total cytoplasmic RNA.

While the early RNAs are copied from a specific portion of the E strand of the DNA, the late RNA is produced from the complementary or L strand. It appears that early transcription is not shut off during this period but that the early sequences constitute only 1 to 5% of the viral RNA during late infection. The L-strand message is produced from the half of the DNA circle opposite that coding for the early transcripts, and the origin of transcription is near the origin of DNA replication, as is the origin of early transcription (see Fig. 1). Like the early transcripts, late mRNA maturation also involves a complex series of "splicing" events to eventually produce several messenger species with both common and unique sequences.

The late gene products which result from the translation of the late message are the viral structural proteins VP1, VP2, and VP3. These viral capsid proteins may be detected as early as 16 h after infection, and the rate of synthesis continues to increase until the capsid proteins compose as much as 10 to 30% of the total cellular protein. These capsid proteins are synthesized from two distinct classes of mRNA. The most abundant class is a 16S mRNA which directs the synthesis of VP1. The other class which sediments at 19S directs the synthesis of both VP2 and VP3. Although it is known that the entire VP3 amino acid sequence is shared with the carboxy terminal region of VP2, it is also clear that VP3 is not a cleavage product derived from VP2 but is in fact synthesized from a unique RNA.

Virus Assembly and Maturation

The assembly and maturation process of the papovaviruses is poorly understood. It is known that replicating DNA rapidly associates with histone proteins, and it is this DNA-protein complex that associates with capsid proteins. There is some evidence for the formation of capsid subunits prior to the association of these structures with the viral DNA; however, this process has not been carefully studied. All attempts at in vitro assembly have been disappointing, although the mixture of DNA-histone complexes with disaggregated capsid structures has yielded some interesting results.

Following assembly the release of the virions appears to be a relatively nonspecific event and is associated with the death of the cells. Even following cell death and rupture some virions still remain associated with the cell debris. There is some evidence that the early stages of virion release are associated with drastic changes in the permeability of the nuclear and cytoplasmic membranes, thereby allowing the virus particles to leave the cell.

Transformation by SV40 and Polyoma Viruses

As described above, cultured cells when transformed in vitro undergo a complex set of changes which have profound effects on their growth. These modifications in their growth and biochemistry may be used to select transformed cells from mixed populations. Selection procedures, such as cloning in soft agar, focus formation, or growth in very low serum, have all been useful in selecting virus-transformed cells from mixed cell populations. Regardless of the system, transformation is an infrequent event, and the use of selective procedures is essential. In one of the most widely studied systems, the transformation of BHK cells by polyomavirus, no more than 5% of the cells in a population will become transformed even at high multiplicities of infection. In the SV40-mouse 3T3 system as many as 40% of the cells may be transformed; however, this frequency requires multiplicities of no less than 10^6 infectious particles per cell.

The low frequency of the transformation event and the high multiplicities required point out one of the essential features of the transformation process, the difficulty in establishing the transformed state. Not only is acquisition of the transformed phenotype an inefficient process, but for every stable transformed cell that emerges from a culture between 10 and 100 cells in that culture transiently acquire the transformed phenotype. These cells are termed *abortively transformed*. The existence of the abortive transformants following the infection of nonpermissive cells helps to point out the various stages in the transformation process. It is clear that during the period of abortive transformation the early viral gene products, most notably the large T antigen, are produced within the cells. However, it would appear that the limiting feature to the stable establishment of the transformed phenotype in these cells is the lack of association between the viral genome and that of the cell. The phenomenon of the abortive transformant gives us one striking piece of evidence, however: the continued expression of the early genes of the virus is necessary for the maintenance of the transformed phenotype.

It is quite clear from the work of several laboratories that at least one round of cell division is required to "fix" the transformed state and convert the abortive transformants into permanently transformed cells. The fixation event occurs in either the S or G2 phase of the cell cycle, and although it has not been established for certain it is very likely that the event that leads to fixation is the integration of the viral DNA into the cellular genome.

The transformation event is not limited to strictly those cell types nonpermissive for the productive growth of the virus. If inactivated virus is used for the transformation event, it is possible to transform even those cells usually permissive for virus replication; however, under these conditions transformation is a very rare event. There have been reports of transformation of permissive cells by both polyoma and SV40. These transformation events seem to have arisen either from the existence in normal virus stocks of defective viral particles or by the generation of random nonpermissive variants in the permissive host cell population.

Genetics of Polyoma and SV40 and Identification of Genes Involved in Transformation

A substantial body of information indicates that the entire SV40 or polyomavirus genome is not required to bring about transformation. Digestion of the viral DNA with restriction endonucleases followed by agarose gel separation of the products produces single fragments capable of transforming cells in culture. Further evidence that only a portion of the viral genome is necessary for the establishment and maintenance of transformation comes from experiments which indicate that conditional mutants in the late proteins cause levels of transformation in nonpermissive cells that are equivalent to the transformation frequencies seen with the wild-type virus. The combination of these genetic studies with the identification of the portion of the physical map of polyoma and SV40 which are capable of transforming cells indicate quite clearly that it is only the early region of these viruses that is required for both the establishment and maintenance of transformation.

Two genes essential for transformation by SV40 have been identified, and the situation is similar, but not identical, in polyoma. The first of these loci is the A gene, which has been identified by the use of temperature-sensitive mutants. These mutants map in slightly different genomic regions in SV40 (middle third) and polyoma (distal third). The function of the A gene polypeptide (large T antigen) has been discussed earlier. The A gene is necessary to initiate transformation, and experiments with temperature-sensitive A gene mutants of SV40 suggest that a functional A gene product is necessary to maintain transformation. The requirement for the continuous presence of the entire A gene polypeptide in polyoma-transformed cells has been questioned. However, a truncated form of the polypeptide is present in transformed cells.

The F gene of SV40 codes for the small T antigen and was identified following the in vitro production of a series of viable deletion (*dl* F) mutants. The mutations map in a portion of the early region which is not part of the A gene by virtue of the removal of that region of the transcript during splicing of the primary transcription product. In lytic infection the viable deletions (*dℓ* F) have virtually no effect, while in the transformed cell these mutants affect several of the factors associated with the transformed phenotype, primarily those associated with anchorage-independent growth. The exact function of the SV40 small T antigen is unknown.

A similar but unique function in polyomavirus is associated with the *hr-t* gene. Like the *dℓ* F mutants of SV40, these too map in the early portion of the genome not included in the A gene product. These mutants were selected because of their altered host range and are only moderately defective in productive infection. They are able to stimulate host DNA synthesis but are totally defective in all other facets of transformation. The *hr-t* functions are probably related to either the small or the middle T antigen.

Although the small T products of the two different viruses are similar, the *hr-t* mutation in polyoma renders it less active in transformation than the *dℓ* F mutants of SV 40. The speculation is that this may be true because the *hr-t* mutation affects both the small and middle T antigens of polyoma.

Nature of Viral DNA in Transformed Cells

Stable transformed cell lines only very rarely show any evidence of infective virus; however, viral DNA is always present and can be demonstrated by a variety of procedures. The amount of viral DNA present in a transformed cell and the organization of that viral DNA is dependent upon the combination of cell type and transforming virus.

Generally, papovavirus DNA is present in transformed cells in numbers between 1 and 10 copies per diploid cell equivalent of DNA. There are a few exceptions, such as polyomavirus-transformed rat cells, which may contain as many as 50 copies per cell. The determination of viral genome copies per cell is generally done by measuring the rate of reannealing of very small amounts of highly radioactive viral DNA in the presence of vast excesses of unlabeled DNA from transformed cells. This technique depends on the fact that the rate of reannealing of a DNA sequence in solution is proportional to the concentration of that sequence in the reaction and is not related to the total DNA concentration. The number of viral genomes per transformed cell can be calculated from the increase in the reannealing rate of the labeled viral probe in the presence of the transformed cell DNA compared with the control DNA. Using this technique it has been clearly established that different lines of transformed cells contain different copy numbers of the transforming virus. In addition, not all segments of the viral genome need be present at the same concentration. Indeed, it is quite clear that several transformed cell lines exist in which the early region sequences have been duplicated in the absence of a comparable duplication of the late segments.

That the DNA of the transforming virus is covalently attached to the high molecular weight cellular DNA has been recognized for many years. Recently, more detailed analysis of this arrangement of viral and cellular DNA has been achieved. The discovery of a large number of restriction endonucleases which cleave high molecular weight DNA at specific sites has provided an efficient mechanism to examine the details of the integration process. These experiments utilize the fragmentation of the transformed cell DNA with one of three types of restriction endonuclease followed by the separation of these fragments by agarose gel electrophoresis. The separated fragments are transfered from the gel to a nitrocellulose filter and hybridized with ^{32}P-radiolabeled viral DNA. Comparison of the migration pattern of the viral sequences in the transformed cell to those of the similarly treated viral DNA alone provides a picture of the nature of the association of the viral DNA with that of the host cell. Figure 2 is a schematic diagram of one such experiment.

First, by using restriction endonucleases with no recognition sites within the viral genome, only the flanking cellular sequences will be cleaved. By counting the number of bands of viral DNA appearing in such digests, the number of separate integration sites of viral DNA can be estimated. The use of enzymes with a single recognition site within the viral genome is useful for determining the number of partial duplications or complete tandem duplications present in the transformed cell DNA. Finally, utilizing nucleases with multiple restriction sites within the viral DNA allows the identification of well-defined regions of the viral DNA and the determination of which of these regions are changed in their migration pattern and therefore associated with host DNA.

When different transformed cell lines are compared following digestion by a restriction enzyme which does not cleave the viral DNA, a comparison of the relative migration of the viral band gives direct evidence of whether the site of integration in each cell line is within the same or a different host restriction fragment. Experiments of this type have clearly shown that there is no specific integration site for the incoming viral DNA but that integration may occur at a number of different locations. Subcloning studies of various single copy cell lines have shown that the chromosomal location of the viral DNA is a stable trait. There are some notable differences in the organization of the integrated DNA in transformed nonpermissive cells compared to transformed semipermissive cells. Transformed semipermissive cells are the only ones in which tandem repeats of two or more viral DNA molecules are seen. As would be expected from such a structure, they are unstable and segregate free circular molecules soon after transformation.

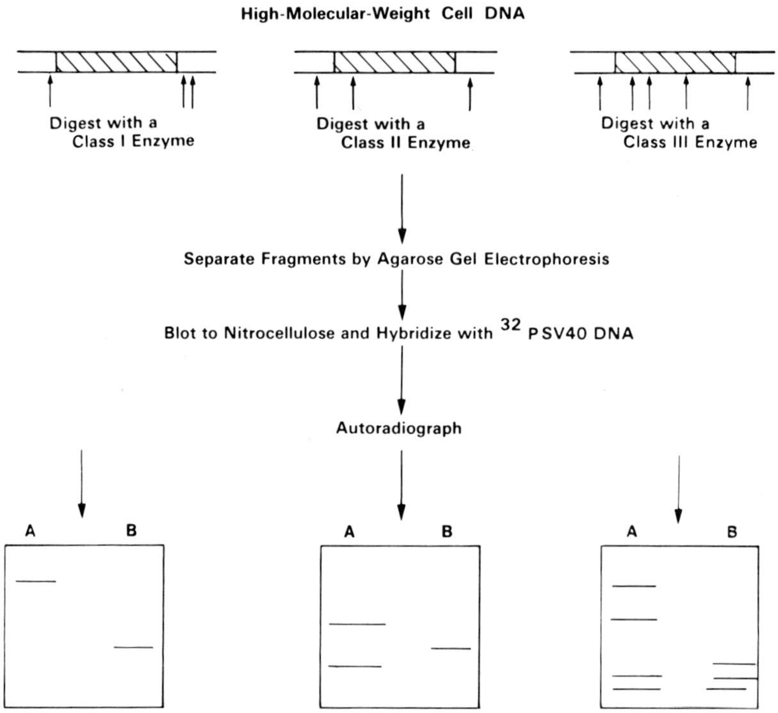

Figure 2 Determination of the arrangement of SV40 sequences in the DNA of transformed cells. Three different classes of restriction endonuclease digestions are shown. Class 1 enzymes do not recognize any site within the viral DNA, class 2 recognize a single site within the virus DNA, and class 3 recognize several viral DNA sites. The digestion of the transformed cell DNA is shown in lane A, while lane B depicts the results expected with mixtures of the nicked circular form of SV40 and normal cell DNA. (Based on the data of Bochan et al. 1976. *Cell* 9:269-287.)

The role of this variable form of integration on the expression of the integrated viral genome remains unclear. As discussed above, the early genes of the papovaviruses are expressed in transformed cells, and at least in the case of SV40 transformants the structure of the mRNA in the transformants is identical to that in early lytic infection. It is unlikely that viral transcription is under host promotor control; however, there is good reason to believe that expression of the viral genes is influenced by the chromatin structure at the site of integration. There is also good evidence for temporal control of virus expression during the cell cycle. Therefore, it appears that although the transcriptional program of the virus is that of early lytic infection and controlled by viral promoters, there are both spatial and temporal constraints imposed by the host cell.

Role of Papovavirus T Antigen in Transformation

Despite an extensive literature regarding the properties of the various T antigens in transformed cells, the mechanism of involvement of these polypeptides in the transformation process has yet to be fully delineated. Additionally, despite their extreme similarities, SV40 and polyoma differ with regard to the role of each T polypeptide. While it is clear that both the large and small T antigens can play a role in transformation by SV40, it is the large T antigen which alone is necessary and sufficient for cell transformation. The available evidence suggests that perhaps two of the activities of SV40 large T polypeptide are essential: the host DNA replication-stimulating activity and the integration of the antigen into the cell membrane. The exact form of this membrane-associated large T antigen is unknown. However, it is clear that only a portion of the large T polypeptide is exposed on the outer cell surface. Two enzymatic activities have been reported associated with highly purified SV40 large T polypeptide, an ATPase and a protein kinase activity. There is some evidence that the protein kinase may not be a part of the large T polypeptide but may instead be associated with a tightly bound protein.

In contrast to the central role for the large T antigen in SV40-transformed cells, polyoma-induced transformation has been achieved with a DNA fragment which lacks the region coding for the carboxy terminal sequences of the large T polypeptide. However, it is clear from several lines of evidence that polyoma middle T is sufficient to establish and maintain transformation. Little information is available about the properties of the middle T antigen. It is known to be phosphorylated and associated with the membrane and has an intrinsic or closely associated protein kinase activity, properties similar to those of the SV40 large T antigen.

The similarities of the SV40 large T and polyoma middle T polypeptides to the *src* gene product of the RNA tumor viruses is striking. The exact role of the protein kinase activities associated with any of these three polypeptides remains at this moment unknown. The role of this enzymatic activity, together with the nature of the ATPase and DNA synthesis stimulating functions, remain some of the most exciting topics in viral oncology.

TRANSFORMATION BY ADENOVIRUSES

The second most widely studied group of transforming DNA viruses are the adenoviruses. This is a large group of agents which, while more complex than the papovaviruses, share many properties with the smaller agents. They are naked capsid viruses with a double-stranded DNA genome. In contrast to the papovaviruses, their DNA is linear in form, with a molecular weight ranging from 20×10^6 to 25×10^6. As might be expected from the larger genome, the adenovirus DNA codes for 20 to 30 polypeptides. At least 15 of these polypeptides are virion structural proteins and form a morphologically complex particle assembled by an appropriately complex process.

Like the papovaviruses, the transcriptional program is under strict temporal regulation during the infection process, with the early genes expressed throughout infection and the late genes expressed only following DNA synthesis. Also in a manner similar to the papovaviruses, nonpermissive cells may become transformed following infection, and these transformed cells contain virus-specific antigens, viral RNA transcripts, and an incomplete copy of the viral DNA covalently attached to host DNA.

Oncogenic Potential of Adenoviruses In Vitro

The adenoviruses were first isolated during the winter and spring of 1952 and 1953 at the National Institutes of Health. These isolates were all obtained from adenoid and tonsillar tissue which had been removed from apparently normal children in the Washington, D.C., area. It was clear at the time that these viruses were previously unidentified agents and were latent in adenoid tissue. Because of the eventual cytopathology in cultured tissue they were initially termed the adenoid degenerating (AD) agents. During that same winter an epidemic of respiratory disease in military recruits at Fort Leonard Wood, Missouri, was associated with influenza virus and another previously unidentified virus. The cytopathic effects of the new isolate from Fort Leonard Wood, called the respiratory illness (RI) agent, were almost identical to those seen with the adenoid degenerating agents isolated at the NIH, and shortly thereafter the two agents were shown to be related in complement fixation and neutralization tests.

In the few years following these initial observations it became clear that this group of agents, eventually termed the adenoviruses, were associated with upper respiratory disease in a number of situations, most notably in military recruit-training camps. During this same period isolates of similar viruses were made from a wide variety of animals, including monkeys, dogs, mice, and cattle. Presently the adenoviruses are categorized into six subgroups based on their natural host species, and the family as a whole has at least 80 members.

The human adenoviruses can be classified into four broad categories based on their clinical significance. The first group, which includes serologic types 3, 4, 7, and 14, is that which is most frequently isolated from patients during epidemic episodes of acute upper respiratory disease and infrequently isolated from long-term adenoid cultures. The second group is best represented by serologic type 8, which is associated with epidemic keratoconjunctivitis and is isolated almost exclusively from patients with this disease. The third group of adenoviruses consists of serologic types 1, 2, 5, and 6, and is usually isolated from long-term adenoid or tonsil cultures of apparently asymptomatic individuals. The fourth group, which includes serologic types 12, 18, and 31, are viruses which are usually isolated from feces of apparently healthy humans, and the incidence of antibody to these viruses is very high. They do not appear to be associated with a specific disease, however, and are not isolated from adenoid or tonsillar tissue.

Because of the problems with upper respiratory disease among military recruits, almost all of whom become infected during recruit training, the preparation of adenoviral vaccines was undertaken in order to reduce the accompanying morbidity. However, while working on these vaccines it was demonstrated that human adenovirus type 12 induced malignant tumors following inoculation of newborn hamsters. The observations on the oncogenicity of the human adenoviruses were extended to a variety of other serologic types, and they were also shown to be oncogenic in other rodent species. However, it was clear that not all human adenoviruses were tumorigenic in rodents, and within the oncogenic types there were variabilities.

Only limited serotypes, for example 12, 18, and 31, are strongly oncogenic and induce tumors rapidly in most inoculated animals. Another group of human adenoviruses, types 3, 7, 8, 14, 21, and 24, have a limited ability to induce tumors and are considered to be weakly oncogenic. The remaining serotypes 1, 2, 4, 5, 6, 9, 10, and 11 show no tumorigenic potential in any rodent species. It has been subsequently shown that the oncogenic potential of each of these three groups is correlated with the base composition of their DNA with the highly oncogenic subgroup having a G + C of 48 to 49% while the weakly oncogenic group has a 49 to 52% G + C and the nononcogenic group have G + C contents of 55 to 60%. There is no evidence that the G + C content of the adenoviruses plays any direct role in the oncogenic process but probably represents the conservation of genes within each group which enhance the oncogenic potential of the virus.

Molecular Biology of Transformation by Human Adenoviruses

The cell-virus interaction seen upon infection with the adenoviruses is very similar to that seen with SV40 and polyoma. For a given serologic type of adenovirus, cultured cells may be either permissive, semipermissive, or nonpermissive, depending on the species of origin of the cells. It is the semipermissive and nonpermissive cells, however, which are most frequently transformed by virus particles.

Experiments with primary cultures of cells of different tissues of rodents have established that the efficiency of transformation depends to a great part on the tissue origin of the cells. For example, rat embryo muscle cells are transformed at one-tenth the frequency that rat embryo brain cells are transformed. Interestingly, cells of the same tissues from different strains of rats are transformed at similar frequencies. In addition to the cell type as a target for transformation, there are a variety of other parameters which affect the transformation process. The multiplicity of infection and the time the cells have been in culture are also significant factors. Once cells are exposed to adenovirus at multiplicities optimum for transformation, it is clear that the transformation process resembles the transformation process previously described for the papovaviruses. The "fixation" of transformation which is seen in both papovaviruses and the RNA tumor viruses is dependent on the mitotic cycle of the infected cell. This requirement for mitosis, which is probably a requirement for the S phase of the cell cycle, is also true of adenovirus transformation. As is the case with the other well-studied tumor viruses, the initial round of host cell DNA synthesis following infection appears to be critical in the establishment of transformation.

Virus-specific gene products are readily detected in adenovirus-induced tumor cells or cells transformed in vitro. However, unlike the situation with papovaviruses, the viral gene products associated with transformed cells vary widely from one transformed clone to another. The basis of this variability in the expression of viral antigens in the cells is due to the fact that the amount of adenoviral DNA present in various transformants is not a constant.

The concentration and specific identification of the sequences of adenoviral DNA present in cell lines transformed by various adenoviruses has been determined in great detail utilizing both the reassociation rate approach described above for the papovaviruses, as well as probing these transformed cells with specific fragments of the adenovirus genome derived by restriction endonuclease cleavage. The results from a variety of laboratories have established the following facts: (1) none of the cell lines that have been examined contain a complete adenovirus genome, and (2) the only sequences present in all transformed cell lines comprise the left-hand 14% of the adenovirus

DNA. Although these data were derived from adenovirus 2-transformed cell lines, it is also clear that a restriction fragment of adenovirus 5 DNA comprising the left-hand 8% of the viral genome will transform rat cells in vitro. While it is clear that the left-hand 8 to 14% of the adenovirus 2 genome is apparently required for transformation, many cell lines which have been examined contain sequences in addition to those required. Similar results have also been obtained with a large number of rodent cell lines transformed by adenovirus 5.

When the restriction endonuclease analysis of transformed cell DNA containing viral sequences was carried out as described above for the papovavirus-transformed cells, it became clear that not only is the size of the adenovirus fragment variable, but likewise there is no unique location for integration in the cellular genome. In addition, these experiments have established the fact that there are no preferred sequences within the adenovirus genome which are necessary as terminal segments for the integration process. In fact, the pattern of integrated adenoviral DNA in transformed cells is often alarmingly simple. Rodent cells transformed by adenovirus 2 or adenovirus 5, which contain small numbers of viral copies, frequently show that all the viral DNA copies are found in the same restriction endonuclease fragment of the cellular DNA. This finding reinforces the previous observation that the tendency of adenovirus-transformed cells is toward aneuploidy, with triploidy being the most common.

Viral mRNA Sequences and Viral Proteins Associated with Transformation

The temporal regulation of transcription in the adenoviruses resembles that seen in lytic infection by the papovaviruses described above. It has been known for some time that the viral messenger RNA molecules present in adenovirus-transformed cells of rodent origin are subsets of the mRNA sequences seen in the early phase of the lytic replication cycle. Some lines which contain large DNA fragments carrying additional information corresponding to early gene products express a larger percentage of their DNA in transformed cells than do those integrated fragments which correspond to a smaller proportion of the early gene sequences and include some of the late gene sequences. This suggests that the mechanism of synthesis and control of the early RNA and transformed cell RNA are very similar. As with the papovaviruses described above, it appears that the transcriptional pattern in transformed cells is similar to that in early lytic infected cells even in the presence of extra adenovirus DNA.

A variety of virus-specific proteins are detectable in adenovirus-induced tumors and transformed cells. These polypeptides have been detected by complement fixation and immunofluorescence using sera from hamsters bearing tumors induced by adenoviruses. While the term *T antigen* has been applied to all the transformed cell antigens that react positively with sera of tumor-bearing animals, it is now clear that a variety of antigens present in transformed cell lines are not necessarily associated with the transformation process. It is also clear that the measurement of key antigens should be limited to those transformed cell lines that have been transformed with the known minimal amount of DNA necessary, so that the expressed antigens are the minimum number known to be required for transformation. While a wide variety of antigens has been detected in transformed cells and a variety of polypeptides has been identified by in vitro translation of mRNA present in transformed cells, only a few are detected in all transformed cells. These include a 58,000 molecular weight polypeptide, a family of four to six polypeptides falling in the range of 38,000 to 50,000, and a 17,000 molecular weight polypeptide. The function of these polypeptides is as yet not understood. None have been purified adequately for biochemical investigations. However, it is clear that there are certain common properties of the adenoviral T antigens present in all cells. Two of these activities are a DNA-binding protein and a membrane-associated antigen. Both of these activities resemble those reported for the large T or middle T antigens of the papovaviruses. There have yet to be adequate studies undertaken to examine whether such enzymatic activities as protein kinase or ATPase are also associated with adenoviral T antigens.

While the in vitro transformation by the adenoviruses and the in vivo oncogenesis by the adenoviruses are accepted fact, several inconsistencies in these viral transformations are yet to be explained. The following questions need to be answered before we understand the tumorigenicity of human adenoviruses and cells transformed by them. (1) Why do transformed cells appear to have the in vitro traits associated with tumorigenic cells and nevertheless fail to produce tumors in immunoincompetent animals? (2) Is the basis for the tumorigenic phenotype the result of a multifactor transformation which involves a second event in adenovirus-transformed cells? We are faced with a dilemma; unlike the situation with the papovavirus-transformed cells, there is not an obvious direct correlation between the trait of transformed cells in vitro and their tumorigenicity in vivo. A second not well understood issue is the role of the various gene products in the transformation process. There seems to be no correlation between the expression of many of the early gene products and the tumorigenicity of the cells. It is clear that more detailed studies of the adenovirus transformation process are necessary before we understand the factors involved in adenoviral oncogenesis.

CELL TRANSFORMATION AND ONCOGENESIS BY HERPES VIRUSES

Herpes viruses are among the most widespread of all the viruses found in nature and have been isolated from humans and almost every species of mammal. Table 1 lists a number of members of the herpesvirus family that have been implicated in oncogenesis in animals or humans or are capable of in vitro transformation of cultured cells. The exact role of any of these in the oncogenic process has not been demonstrated. However, it is clear that in one instance, Marek's disease of fowl, immunization against herpesvirus prevents the disease.

Viruses are placed in the herpes group based on several criteria, most notably their morphology in the electron microscope and the presence of a large double-stranded DNA molecule in the virion. Herpesviruses contain four major morphologic elements: (1) an electron-dense core which consists primarily of the compacted virion DNA; (2) an icosahedral capsid surrounding the core; (3) electron-dense amorphous material, the tegument, which lies between the nucleocapsid, and (4) the outer envelope. The outer envelope consists of cell membranes which have been modified during virus infection.

All members of the herpes group contain large double-stranded DNA molecules which range from 80×10^6 to 150×10^6 in molecular weight. The most thoroughly characterized of the herpes group are herpes simplex virus types 1 and 2 (HSV 1 and HSV 2) which contain linear DNA molecules of approximately 100×10^6 molecular weight. The DNA of these viruses is unusual in that it has an extremely high G + C content of 67% for type 1 and 69% for type 2.

Transformation by Herpes Simplex

Unlike the situation with the papovaviruses and the adenoviruses, there is no well-characterized pattern of cultured cells that are permissive, semipermissive, or nonpermissive for HSV. The herpes simplex viruses are invariably some of the most cytopathic of mammalian viruses, and the invariable result of infection is cell death. Therefore, studies on the transformation of cells in culture by HSV have utilized viruses that have been inactivated by various means. Most studies on herpes-transformed cells have examined cells that have been transformed by UV-inactivated virus, virus resulting from photodynamic inactivation, cells infected at the nonpermissive temperature with temperature-sensitive viral mutants, or cells transfected with restriction fragments of herpes simplex virus genome.

Transformation by HSV falls into two categories: (1) those transformants which have been selected for the expression of a viral gene, thymidine kinase, referred to as biochemical transformants, and (2) cells which have been selected for those biologic characteristics associated with oncogenic transformation, that is, altered cell growth. Our understanding of the molecular biology of the transformation process by herpes virus is not nearly as complete as our understanding of the similar process by the papovaviruses or adenoviruses. It is clear that all cells transformed by HSV, whether they are biochemical or morphologic transformants, contain a fragment of the herpes viral genome which is probably integrated into the host DNA. There is little information about the size of the required viral DNA fragment necessary for this transformation process, and it is generally accepted that the amount of DNA initially observed in transformed cells is frequently modified during the continued growth of the cells in culture. The results of these studies suggests that only a very tiny fragment of perhaps no more than 2×10^6 molecular weight is necessary for the continued maintenance of the transformed phenotype in herpes-infected cells. There is no consensus on the amount of information necessary to establish transformation.

Gene expression in herpes-transformed cells has been characterized in several different transformants and has been shown to vary from isolate to isolate. Because most biochemically transformed cells are continuously cultured under selective conditions, the expression of the thymidine kinase gene is maintained during culture, whereas the expression of other associated herpes viral genes varies with the cell isolate. The expression of herpesvirus antigens in morphologically transformed cells likewise varies from cell line to cell line, and it is clear that a variety of HSV glycoproteins are expressed in these cells. The unfortunate situation with HSV-transformed cells is that the expression of the viral genes is highly variable from one cell line to another, and therefore the minimum number of required viral gene products has yet to be determined. Clearly, the analysis of gene expression in cell culture lines transformed with the minimum necessary viral DNA fragment will help yield the necessary information about the required viral gene products. The temporal regulation of productive infection by HSV is far more complex than that of other DNA viruses. The expression of herpes viral genes cannot be simply described as early and late functions controlled by the onset of viral DNA synthesis, but instead the transcriptional program falls into a series of classes which show differential expression during the replication cycle. For that reason it is very difficult to examine transformed cells and categorize the antigen present as being associated with HSV early or late functions.

HSV and Human Cancer

Antibodies to HSV 2 have been observed at a higher frequency in women with cervical cancer than in women in control groups. These observations stimulated the examination of HSV as both a transforming virus in vitro, as described above, and likewise the search of human tumors for the presence of viral antigens and HSV DNA. With one exception, experiments examining human tumors for the presence of HSV DNA, utilizing the reassociation rate kinetic approach described above for the quantitation of viral DNA in papovavirus-transformed cells, have yielded negative results regarding the presence of HSV DNA in these human cancers. However, recent work utilizing an in situ hybridization procedure in which tumor cells are examined for the presence of viral RNA transcripts, have suggested that many of these human cancers contain RNA complementary to HSV DNA. These experiments, when coupled with the observations from two different laboratories that a specific restriction fragment of HSV DNA is capable of transforming cells in culture, have stimulated a new round of interest in the association of HSV with human cervical carcinoma. At present the subject of the association of HSV 2 with cervical cancer remains controversial. However, the evidence cited above suggesting that there may be some association between the two has continued to stimulate further study, and soon this question should be resolved.

Oncogenesis and the Epstein-Barr Virus

Between 1958 and 1962, Burkitt described a childhood lymphoma common in equatorial East Africa and which has subsequently been found in other parts of the world. This lymphoma occurs in areas where there is a high incidence of arthropod-borne infection, which led Burkitt to hypothesize that it was of infectious etiology and associated with arthropod-borne disease. Epstein and his collaborators, working with cultured cells derived from Burkitt's lymphomas, detected a herpes-like virus which has subsequently been called Epstein-Barr virus (EBV). It has been established that antibodies to EBV antigens occur in more than 80% of healthy adults around the world and are present in virtually every patient with a Burkitt's tumor. Studies to test the idea that EBV is the etiologic agent of Burkitt's lymphoma have established clearly several facts: (1) EBV is the etiologic agent of infectious mononucleosis, and (2) EBV has a very close relationship to both Burkitt's lymphoma and to nasopharyngeal carcinoma in the far East. However, the results of these studies clearly have suggested that not only is the EBV involved in these tumors but likewise some other factors must be cooperative events in the induction of these cancers.

Epstein-Barr virus is able to induce tumors in experimental primates as well as to transform cells in culture. Despite its ubiquitous nature, the EBV has a very limited host range, and only a few cultured cells have been shown to have the appropriate receptor sites for infection. These receptor sites can be demonstrated only on B lymphocytes of humans and primates and on nasopharyngeal epithelial cells. When human lymphocytes are placed in culture together with EBV, there appear, after a few days, cells which begin to multiply and which eventually develop into continuous cell lines. This transformation process can be blocked by pretreatment of the virus pool with antiviral serum.

The host-virus relationship in EBV-transformed cells is one of great complexity. Few if any cells in a culture are actively replicating the virus, although some cells derived from Burkitt's tumors can be induced to produce virus by treatment with inducing agents. Cells actively producing EBV appear to go on to cell death upon the release of infectious virus; however, fewer than 20% of the cells in any induced culture are virus producers.

Nucleic acid reassociation kinetic studies have shown that Burkitt's tumor cells in culture, as well as transformed lymphocytes, contain copies of the EBV DNA ranging from a low of two to three copies per cell up to cell lines containing as many as 200 copies per cell. Most transformed cell lines established from tissues of Burkitt's lymphomas contain from 20 to 60 EBV genomes per cell. There is evidence that at least in some cell lines the viral DNA is integrated into the host genome; however, it is clear that many cell lines have nonchromosomal EBV DNA that can be isolated as circular cytoplasmic molecules.

A variety of identifiable EBV antigens are present in transformed cells, although these antigens are variable from one cell line to another. The most commonly identified EBV-associated antigen in transformed cells is the Epstein-Barr nuclear antigen (EBNA), which was first detected by complement fixation in nonproducing cell lines. EBNA can be shown to reside in the nucleus of the cell and is associated with cellular chromatin. It is present in all cells infected with EBV regradless of the source. EBNA is a basic protein of a molecular weight of approximately 150,000 to 200,000 and appears to have a DNA-binding activity. Its activity and presence in the nucleus of cells more closely resembles the SV40 large T antigen than any of the other EBV-associated antigenic molecules. A variety of other EBV-associated antigens are found in transformed cells. These are the lymphocyte-detected membrane antigen (LYDMA) and the various other membrane antigens detected by immunofluorescence. The role of these various antigenic components in the transformation process remains a mystery.

Relationship Between EBV and Human Cancer

Despite the apparent uniform presence of anti-EBV antibody in patients with Burkitt's lymphomas and nasopharyngeal carcinoma, it has yet to be fully established that this virus is in fact the etiologic agent of these diseases. Recent large-scale seroepidemiologic studies sponsored by the World Health Organization have shown that virtually 100% of children in the West Nile district of Uganda have acquired antibody to EBV by the age of 4 years, but only about 1 in 1000 or fewer of these children go on to develop Burkitt's lymphoma. Children with unusually high EBV antibody titers appear to be at a 30-fold increased risk for the development of Burkitt's lymphoma when compared to the average titer of the total population of children. These observations, together with earlier seroepidemiologic studies, lead to the conclusion that the tumor cannot be a result of the direct infection of the individual by EBV. The results seem clear that some other second factor must be superimposed on the infection by EBV in order to induce tumor development. If infection by EBV is necessary for the development of this tumor, it is certainly not sufficient for that development. The reasons it is suspected that EBV may have an etiologic role in Burkitt's lymphoma can be summarized as follows:

1. EBV has oncogenic potential both in vivo and in vitro, and the target cells of this oncogenesis are B lymphocytes, as are cells in Burkitt's tumors.
2. Burkitt's lymphoma is a monoclonal disease, and therefore the cell that gives rise to the tumor must have been infected prior to the onset of the uncontrolled multiplication.
3. The cells of the Burkitt's tumors that arise in Africa, as well as many of those that arise elsewhere, contain EBV DNA and EBV antigens.

In contrast to these factors, the following observations raise questions as to the true role of the virus in the tumor induction process. First, a small minority of the Burkitt's tumors in Africa, as well as the majority of the Burkitt's tumors in the rest of the world, do not contain EBV DNA or EBV antigens. Second, in the African tumor belt, the vast majority of children are infected with EBV by age 3 years but never develop the tumor. It is clear that other environmental or genetic factors are necessary to be superimposed on the presence of EBV for the induction of this tumor.

Nasopharyngeal carcinoma is prevalent in certain isolated regions of the world, most frequently in Southeast Asia, where it is particularly prevalent in Cantonese Chinese. Patients with this tumor have elevated titers of antibody against the viral capsid and membrane antigens of EBV, as well as neutralizing antibody. Using DNA-DNA hybridization it has been demonstrated that there is EBV DNA in biopsies from virtually every one of these tumors. Because of the difficulties in separating the potential casual relationship of the virus with the tumors due to the spread of the virus from the patient's lymphocytes to the tumor cells, which have been recognized to have the appropriate EBV receptors, the relationship between EBV and nasopharyngeal cancer is not any clearer than the relationship between EBV and Burkitt's lymphoma.

Clearly, the Epstein-Barr virus is more closely associated with human malignancies than any other virus known today. However, it is also very clear that the relationship between EBV and human malignancies is one which is far from a simple causal relationship. Studies of a variety of viruses of primates which appear to be closely related to the EBV of humans may perhaps reveal the mechanisms of the relationship between the virus and the tumors and allow a clearer definition of the role of this virus in human tumors.

Marek's Disease Virus and Tumors in Fowl

The chicken herpesvirus, Marek's disease virus (MDV), causes an infection of the epithelial feather follicles which is productive for virus, while at the same time the virus abortively infects the T lymphocytes of the chicken and leads to neoplastic transformation. The disease is a generalized lymphomatosis and involves lymphocytic invasion of a variety of organs, including nerve trunks, and frequently results in paralysis of the birds. The disease may lead to high rates of mortality in affected flocks. The virus was first isolated by co-cultivation of infected tumor cells with embryonic fibroblasts, and subsequently it has been shown that the only site of productive replication in the chicken is the feather follicle epithelium.

Antigenic structural proteins of the virus are not detected in cells of biopsies, nor has any other viral antigen been detected in these cells. It is clear, however, that the viral genome is carried by the cells in the tumors and by cultivated tumor cells. In the tumor cells there appear to be somewhere between 20 and 80 viral genome equivalents per cell.

Initial immunization attempts utilized a live attenuated strain of MDV as a vaccine. More recently, however, a herpesvirus of turkeys which is very closely related antigenically to MDV has been used as a viral vaccine strain. This vaccination has proven extremely effective for the elimination of Marek's disease from infected flocks. These immunization results have led to the conclusion that there is a direct relationship between MDV and the development of the lymphoma. Whether MDV is the single and only etiologic factor necessary for this development, however, has not been fully established.

SUGGESTED READING

Tooze, J. (ed.). *Molecular Biology of Tumor Viruses. Part 2. DNA Tumor Viruses*. Cold Spring Harbor Laboratory, Cold Spring Harbor, New York, 1980.

Index